EYES ON THE SKIES

400 Years of Telescopic Discovery

WILEY-VCH

EYES ON THE SKIES

400 Years of Telscopic Discovery

The Authors

Govert Schilling

Amersfoort, The Netherlands

Lars Lindberg Christensen

ESA/Hubble

ESO/ESA/ST-ECF

Garching, Germany

Design and Layout

Martin Kornmesser &

Nuno Marques

ESA/Hubble

ESO/ESA/ST-ECF

Garching, Germany

STAR-FORMING REGION S106
(INSIDE FRONT AND INSIDE BACK COVER)

The Subaru Telescope on Mauna Kea produced this deep infrared image of the star-forming region S106. S106 is at a distance of approximately 2000 light-years from Earth. The hourglass appearance of S106 is the result of the way material is flowing outwards from the central star. This infrared image is very sharp and reveals subtle details like ripples inside the nebula.

PINWHEEL GALAXY (PAGE 3)

The Pinwheel Galaxy's giant spiral disc of stars, dust and gas is 170 000 light-years across or nearly twice the diameter of our Milky Way. The galaxy is estimated to contain at least one trillion stars. Approximately 100 billion of these stars alone might be like our Sun in terms of temperature and lifetime. The high resolution of the *Hubble Space Telescope* reveals millions of the galaxy's individual stars in this image.

All books published by Wiley-VCH are carefully produced. Nevertheless, authors, editors, and publisher do not warrant the information contained in these books, including this book, to be free of errors. Readers are advised to keep in mind that statements, data, illustrations, procedural details or other items may inadvertently be inaccurate.

Library of Congress Card No.:
applied for

British Library Cataloguing-in-Publication Data
A catalogue record for this book is available from the British Library. Bibliographic information published by the Deutsche Nationalbibliothek

The Deutsche Nationalbibliothek lists this publication in the Deutsche Nationalbibliografie; detailed bibliographic data are available on the Internet at <http://dnb.d-nb.de>.

© 2009 WILEY-VCH Verlag GmbH & Co. KGaA, Weinheim

Printed in the Federal Republic of Germany
Printed on acid-free paper

Composition: Hagedorn Kommunikation GmbH, Viernheim
Printing: betz-druck GmbH, Darmstadt
Bookbinding: Litges & Dopf GmbH, Heppenheim

ISBN: 978-3-527-40865-8

STAR TRAILS OVER GEMINI NORTH

Star trails majestically arch above the Gemini North Telescope in this long exposure, made on Mauna Kea. Light from the setting Moon is reflected from the right of the dome, while twilight provides a faint lingering glow on the left side. The centre of the dome shows the glow of a small red flashlight.

TABLE OF CONTENTS

FOREWORD

We believe that even our most remote ancestors looked up with wonder and awe at the night sky. But, 400 years ago, something entirely new happened: Galileo turned a homemade arrangement of magnifying glasses to the skies, taking advantage of information on advances in spyglasses made elsewhere in Europe. So Galileo was the first to see amazing phenomena: the mountains on the Moon, the phases of the planet Venus, the satellites of Jupiter, spots on the Sun... But the essential step taken by Galileo, the most important for us, as astronomers, is that he immediately tried to understand the meaning of what he saw, to translate the beautiful images into facts about the Earth and its position and motion with respect to the Sun; he recognised the Moon as a body akin to the Earth, and he realised that Jupiter and its satellites formed a miniature Solar System.

This happened just 400 years ago. Since then astronomers have been following in Galileo's footsteps, constantly improving telescopes and instruments, and trying to make sense of it all. The progress has been fantastic! Today, there are many optical telescopes with diameters in excess of eight metres all over the world, supplemented on the ground by radio telescopes and novel detectors of ultra high-energy particles and photons, while satellites pick up other kinds of radiation from the Universe. Space is expensive, but offers the ideal conditions for observations, even in the optical and the infrared, as can be seen from the unique results obtained by the modestly sized Hubble Space Telescope, which complement those of the giant Earth-bound telescopes. These new observations, and the interpretations and theories they have fostered, have not only revolutionised our view of the Solar System, but have turned the entire Universe, its origin, its evolution, the history of its components, gas clouds, galaxies and galaxy clusters, stars, planets in our Solar System and elsewhere, into objects of scientific study.

So, in 2009, in the International Year of Astronomy, we are celebrating Galileo's legacy and all the discoveries that have taken place in the intervening years, as well as the explosion of knowledge that we are witnessing now, made possible by new technologies. This book, written by two experienced and talented astronomy communicators, illustrates beautifully the saga of the telescope over 400 years, and the prodigious advances that have been made in understanding the Universe.

C. Cesarsky

Catherine Cesarsky
President of the International Astronomical Union & Director of Research, DSM/CEA-Saclay, France

1 NEW VIEWS OF THE SKIES

The telescope is astronomy's miracle worker. It reveals faint stars and nebulae and magnifies distant objects. Telescopes take astronomers on a journey to the distant reaches of the Universe, where sparkling galaxies adorn the darkness of the void. But they also serve as time machines, providing scientists with a view of the earliest cosmic eras. No other single instrument has done so much for our view of mankind's place in time and space. Astronomy would barely rate as a science without the telescope. Four hundred years ago the early pioneers began a journey that led from the chance alignment of two simple lenses to today's complex space-based observatories and massive mountaintop mirrors.

Four centuries ago a man walked out into the fields near his home in Padua and pointed his homemade telescope at the Moon, the planets and the stars. Astronomy would never be the same again. The date was Thursday, 30 November 1609. The man was the Tuscan physicist and astronomer Galileo Galilei. He may not have realised it at the time, but that night he started a scientific revolution of cosmic proportions. To commemorate Galileo's first observations of the heavens with a telescope the United Nations and the International Astronomical Union have declared 2009 to be the International Year of Astronomy.

For thousands of years the human eye was the only instrument available to observe the Universe. The invention of the telescope changed that. Now astronomers assemble giant mirrors on remote mountaintops to look out through the thinnest layers of the clearest, stillest atmosphere to catch faint signals from some of the farthest and oldest objects known. Radio telescopes collect faint chirps and whispers from outer space. Scientists have even launched telescopes into Earth orbit, high above the distorting effects of our atmosphere. And the view has been breathtaking.

Galileo didn't invent the telescope, and its exact origin is still controversial. The oldest existing documents to mention the telescope attribute its invention to the Dutch spectacle maker, Hans Lipperhey (also known as Lippershey) in the early 17th century. Tinkering away, Lipperhey found that placing a convex lens at one end of a cardboard tube and a concave lens at the other allowed him to magnify distant objects. The telescope was born!

> *" The true origin of the telescope remains shrouded in mystery "*

As far as we know, Lipperhey never looked at the stars through his telescope — he believed that his invention would mainly serve seafarers and soldiers. Lipperhey was from Middelburg, a large trading city in the fledgling Dutch Republic, then at war with Spain. In October 1608 Lipperhey demonstrated the telescope to Prince Maurits of the Netherlands, who was able to read the time on the church clock in Delft, from a tower in The Hague eight kilometres away. The new spyglass could reveal enemy ships and troops too distant to be seen by the unaided eye. A useful invention indeed. However, the Dutch government did not grant Lipperhey a patent, since other merchants, notably Lipperhey's competitor Zacharias Janssen, also claimed the invention and might actually have built the first telescope around 1604. The dispute has never been settled. The true origin of the telescope remains shrouded in mystery.

Who was the first?

Hans Lipperhey's 1608 patent application for a "tube to see far" is the oldest existent document related to the invention of the telescope. But some historians think that his Middelburg competitor Zacharias Janssen was the first to build a working version of the instrument. The truth may never be known.

Lipperhey (above) demonstrated his telescope to Prince Maurits in early October 1608, and filed a patent application with the Dutch States-General. But word about the telescope had already been around for quite some time. In mid-September, Lipperhey received a visit from Jacob Metius from the town of Alkmaar, who had been making his own experiments and wanted to learn more about the Middelburg invention. Metius's interest may well have been the main reason for Lipperhey to think about making a patent application. Metius himself also applied for a patent, just a few days after Lipperhey's demonstration. At this time Lipperhey's fellow spectacle maker Zacharias Janssen was out of the country on business, but when Janssen returned to Middelburg in mid-October he testified that he had been the first to build telescopes. As a result of the dispute, Lipperhey's patent was never granted, although the States-General did pay him a large sum of money for the construction of a binocular telescope, which he delivered in December of that year.

Is there any truth in Janssen's story? Willem Boreel, the Dutch envoy in Paris, who researched the issue forty years later on behalf of Louis XIV's personal physician — and amateur astronomer — Pierre Borel, concluded that Janssen had been the first to build telescopes. Janssen's claim also fits in with a story discovered only in the early 20th century. In 1634, a few years after Janssen's death, the Dutch scientist Isaac Beeckman paid a visit to Middelburg to learn the art of grinding lenses from Janssen's son, Johannes Zachariassen. In his personal diaries, Beeckman recounts Zachariassen's story that his father built the first telescope in 1604.

" It was time to train the telescope on the heavens "

In Padua, Galileo Galilei heard about the telescope through a French colleague. Galileo was the greatest scientist of his time: he studied falling objects, devised the laws of motion, and challenged the age-old views of Greek philosopher Aristotle, establishing the foundations of the modern scientific method. Galileo was also a strong supporter of the new worldview of the Polish astronomer Nicolaus Copernicus, who proposed that the Earth orbited the Sun rather than the other way round.

Galileo soon built his own instruments based on the reports of the Dutch invention. They were of much better quality, with a wider field of view, a clearer image and a higher magnification. *"Finally,"* he wrote,

"sparing neither labour nor expenses, I succeeded in constructing for myself so excellent an instrument that objects seen by means of it appeared nearly one thousand times larger than when regarded with our natural vision."

It was time to train the telescope on the heavens.

Was the telescope invented in England?

According to some historians, the English mathematician and surveyor Leonard Digges built the first telescope decades before either Lipperhey or Galileo. Writing in 1571, his son Thomas described his father's experiments with "perspective glasses". Moreover, Digges's contemporary William Bourne wrote: "For to see any small thing a great distance from you, it requireth the ayde of two glasses." The English astronomer Thomas Harriot is also reported to have shown "perspective glasses" to Native Americans during his trip to Virginia in 1586. However, there is no convincing proof of the existence of an Elizabethan telescope — Digges's "perspective glasses" may simply have been hubcap-like convex mirrors.

" A landscape of craters, mountains, and valleys. A world like our own."

GALILEO'S MOON DRAWINGS FROM 1610

On 30 November 1609, Galileo made his first telescopic sketches of the Moon, showing mountains and craters. Although Thomas Harriot in England had already observed the Moon through a telescope in the summer of 1609, Galileo was the first to publish his Moon drawings, and to give detailed descriptions of what he had seen.

On 30 November 1609, just outside his home in Padua, Galileo aimed his telescope at the Moon, recording his first glimpse of extraterrestrial terrain:

> *"I have been led to the opinion and conviction that the surface of the Moon is not smooth, uniform, and precisely spherical as a great number of philosophers believe it to be, but is uneven, rough, and full of cavities and prominences, being not unlike the face of the Earth."*

A landscape of craters, mountains, and valleys. A world like our own.

Who was the first to use the telescope for astronomy?

Another somewhat contentious topic is the question of who started using the telescope for astronomical purposes. Was it the English astronomer Thomas Harriot or the Italian astronomer Galileo Galilei? Thomas Harriot was a scientist and astronomer living in Oxford. He was at one point a cartographer on an expedition organised by Sir Walter Raleigh. It is claimed by some that Harriot observed and sketched the Moon through a telescope 5 August 1609, months before Galileo is known to have done so on 30 November 1609. This may very well be true, but the evidence is hard to come by as Harriot did not publish his results — as with all his scientific discoveries. Though hardly as famous as Galileo's, Hariott's observations of sunspots from 3 December 1610 do seem to be the first such observations of this phenomenon.

" Not everything revolves around the Earth, as the Greeks had always believed "

A few weeks later, in January 1610, Galileo looked at Jupiter. Close to the planet, which itself was clearly a small sphere, he saw four "stars" that changed position from one day to the next. This slow, celestial ballet of satellites orbiting another planet convinced Galileo that Copernicus was right: not everything revolves around the Earth, as the Greeks had always believed.

And there was more. The phases of Venus: just like our moon, this brilliant planet, visible in the morning or evening sky, waxes and wanes from crescent to full and back again — as a result of its orbit closer to the Sun. On either side of Saturn, Galileo saw strange appendages, looking like the handles of a Roman amphora that grew larger and then shrank back over the years. Risking the loss of his eyesight, he also observed dark spots

GALILEO'S DRAWINGS OF VENUS

According to the old geocentric worldview of the Greeks, the planet Venus could never be farther from the Earth than the Sun is. However, Galileo's observations revealed that Venus goes through phases, just like the Moon. Since the planet shows a gibbous phase (right) when it is close to the Sun in the sky, it must be behind the Sun from our perspective. This was the most convincing telescopic "proof" of Copernicus' heliocentric theory.

on the blindingly bright surface of the Sun. And then there were the stars: thousands of stars, maybe even millions, too faint to be seen by the naked eye. It was as if mankind had discarded its blindfold. There was a whole Universe out there to discover.

News of the telescope spread across Europe like wildfire. In Prague, at the court of Emperor Rudolph II, Johannes Kepler, a great admirer of Galileo, improved the design of the instrument. In Antwerp, the Dutch cartographer Michael van Langren used a telescope to produce the first reliable map of the Moon, showing what he believed to be continents and oceans. And Johannes Hevelius, a wealthy brewer and amateur astronomer in Poland, built huge telescopes on his rooftop observatory in Danzig.

Astronomy before the telescope

What did astronomers do before the telescope was invented? They saw the same as everyone else: the changing phases of the moon, wandering planets, solar eclipses, comets and shooting stars, and the occasional supernova, but they were still scientists. So they recorded what they saw and devised philosophies in which all these celestial phenomena fit together. The Greeks believed that the Earth occupied the centre of the Universe, with the Sun, Moon and planets moving around it on crystal spheres. Copernicus placed the Sun at the centre with the planets revolving about it, but kept the fixed stars on a crystal sphere at the rim of the Solar System. It wasn't until the late 16th century that scientists first thought of stars as other suns, strewn throughout three-dimensional space.

NEWTON'S REFLECTING TELESCOPE

Only 15 centimetres long, the first reflecting telescope built by Isaac Newton could magnify by about 40 times — more than a two metre long refracting telescope of its day. Newton's new design minimised the problem of chromatic aberration — colour defects common to refractors. However, due to problems with accurately grinding the metal mirror, Newton's first reflector, a replica of which is seen here, actually caused more image distortions than other contemporary telescopes. As a result, more than a century passed before reflecting telescopes became popular among astronomers.

But the best instruments of the time were probably constructed by Christiaan Huygens in Holland, the brilliant son of a wealthy Dutch poet and diplomat. With his brother Constantijn, Huygens ground lenses of very high quality, some of which are still preserved. In 1655 Huygens discovered Titan, the largest moon of Saturn. A year later, his observations revealed the true nature of Saturn's ring system — something Galileo had never understood. And last but not least: Huygens saw dark markings and bright polar caps on Mars. Could there be life on this remote, alien world? The question has held the public imagination ever since.

The earliest telescopes used a convex lens to collect and focus starlight. These refracting telescopes suffer from chromatic aberration: different colours are refracted in slightly different ways. As a result, the star images have tiny rainbow haloes. But in a reflecting telescope — first built by Niccolò Zucchi and improved by Isaac Newton 1668 — it's all done with concave mirrors, and there are no colour defects. Unfortunately, in Newton's time the mirrors were made from polished copper or tin and not of particularly high quality.

" Herschel must surely have felt like a child let loose in a sweetshop "

In the late eighteenth century, the largest telescope mirrors in the world were cast by William Herschel, a German-born organist turned astronomer, who worked with his younger sister Caroline. In their house in Bath, England, the Herschels poured red-hot molten metal into a mould and then polished the surface so that it would reflect starlight. The largest of Herschel's wooden telescopes was so enormous — its mirror measured 1.2 metres in diameter — that he needed four servants to operate the wheels, ropes and pulleys used to allow the telescope to track the apparent motion of the sky that results from the Earth's rotation around its axis. Herschel scanned the heavens systematically and catalogued hundreds of nebulae and binary stars. As the first one to do so, he must surely have felt like a child let loose in a sweetshop.

The first planet-spotters

The planet Uranus was discovered in 1781 by William Herschel. Neptune was found in 1846 by Johann Galle, on the basis of predictions made by Urbain Le Verrier. Both planets had been observed before by famous astronomers, but had gone unrecognised. The English astronomer John Flamsteed catalogued Uranus as Star Number 34 in the constellation of Taurus the Bull in 1690. Galileo spotted Neptune even earlier, in 1612, when the planet was close to Jupiter in the sky, but he mistook it for a background star.

> "*The largest of Herschel's wooden telescopes was so enormous that he needed four servants to operate the wheels, ropes and pulleys*"

HERSCHEL'S LARGEST TELESCOPE

With a mirror measuring 1.2 metres in diameter, William Herschel's largest telescope was an unwieldy instrument. It remained the biggest telescope in the world until the mid-eighteenth century. In the 1830s Herschel's son John spent a few years in South Africa where he set up a similar, but smaller, telescope to study the southern skies.

" On the occasional clear, moonless night, Lord Rosse sat at the eyepiece, and sailed on a journey through the Universe "

HERSCHEL'S DRAWING OF THE MILKY WAY

William Herschel worked at the cutting edge of astronomical debate in his era and spent much of his time attempting to unravel the mysteries of the Universe. Here we see William Herschel's drawing of the Milky Way, which was included in his paper on the construction of the night sky. The drawing demonstrates how an individual located at a point in the middle of a thin and flat layer of stars will see them.

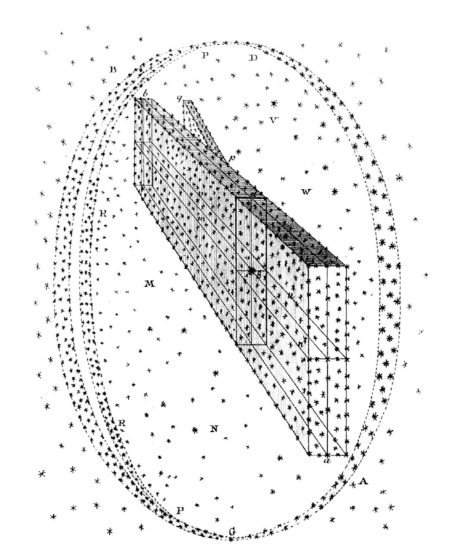

From his meticulous star counts, Herschel discovered that the Milky Way must be a flat disc. He even measured the motion of our Solar System through space. And on 13 March 1781, Herschel found a new planet, Uranus — a tiny, blue-green sphere, slowly gliding against the backdrop of distant stars, way out beyond the orbit of Saturn. In one instant, Herschel had doubled the size of the Solar System. It was more than two hundred years before NASA's *Voyager 2* spacecraft gave astronomers their first close-up look at this remote world.

LORD ROSSE'S LEVIATHAN

Erected between 20-metre high walls and resting on a giant ball bearing, Lord Rosse's "Leviathan of Parsonstown" was the world's biggest telescope from its completion in 1847 until the construction of the 2.5-metre Hooker Telescope at Mount Wilson, California, in 1917. The telescope could move up and down, but only a little to the right and left, enabling observations of a big swath of the sky above the southern horizon.

In the mid-1840s, in the lush and fertile countryside of central Ireland, William Parsons, the wealthy 3rd Earl of Rosse, built the largest telescope of the nineteenth century, known as the "Leviathan of Parsonstown". Its 3.5-ton metal mirror was a staggering 1.8 metres across and its 18-metre tube was erected between two masonry walls that towered over 20 metres high. On the occasional clear, moonless night, Lord Rosse sat at the eyepiece, and sailed on a journey through the Universe fascinated by the fantastic shapes of distant nebulae and remote "island universes".

" *The telescope had become our vessel to explore the Universe* "

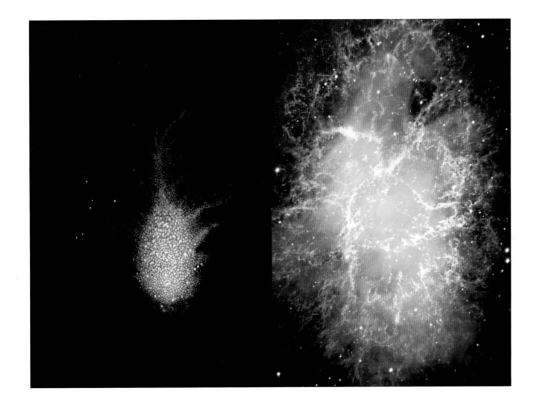

THE CRAB NEBULA

The Crab Nebula takes its name from the tentacle-like protrusions, visible in Lord Rosse's drawing (left). As the Very Large Telescope image at right shows the nebula is an expanding cloud of gas energised by the fierce radiation of a central neutron star. This is all that remains of the supernova explosion from the year 1054 that originally produced the nebula.

He recorded his observations in drawings and sketches, showing new details of some of today's most familiar astronomical objects: the Orion Nebula, now known to be a stellar nursery where new suns are born, and the mysterious Crab Nebula, the remnant of a supernova explosion. Lord Rosse's drawings date from an era before astronomical photography and were the first to show the majestic spiral shape of the Whirlpool Nebula. We now know that the Whirlpool is another galaxy, with intricate clouds of dark dust and glowing gas, billions of individual stars, and — who knows — maybe even planets like Earth.

The telescope had become our vessel to explore the Universe.

THE WHIRLPOOL GALAXY

Lord Rosse made this illustration using his giant reflecting telescope, known as the Leviathan. With this instrument he was the first to recognise the spiral shape of some fifteen nebulae, including M51, or the Whirlpool Galaxy. This illustration was taken from a paper written by Rosse, published in 1850 and entitled "Observations on the Nebulae". Because of its striking appearance, Lord Rosse sometimes called the Whirlpool Galaxy the Question Mark Nebula. His 1845 drawing (right) compares well with the Hubble Space Telescope photo (left). The "point" of the question mark is actually a small, interacting companion galaxy.

BIGGER IS BETTER

In their quest for ever-fainter objects and finer detail, astronomers have always demanded bigger telescopes. Scientific vision, technical nerve and personal perseverance led to the giant observatories of the early twentieth century. Located on remote peaks and protected beneath majestic domes, these awe-inspiring instruments have revealed an expanding and evolving Universe, populated by a stunning variety of galaxies and nuclear-powered stars that produced the elements in our bodies. A few decades ago the 5-metre Hale Telescope on Palomar Mountain seemed to be the ultimate telescope. But was it?

" How to do better? Think mirrors. "

THE YERKES REFRACTOR

Ground by American telescope builders Alvan Clark & Sons, the primary lens of the Yerkes Observatory's Great Refractor, in Williams Bay, Wisconsin, has a diameter of 40 inches (101.6 centimetres). The lens was completed in 1895 and is still the largest ever made.

At night your eyes adapt to the darkness and your pupils widen to collect more light. As a result, you can see dimmer objects and fainter stars. Suppose you had pupils one metre across — that would give you supernatural eyesight, showing stars 25 000 times fainter than those we can see with our eyes! This is where telescopes come in. A telescope is like a funnel. The lens or mirror collects starlight over its whole surface and concentrates the light into a narrow beam that enters your eye. The bigger the lens or mirror, the fainter objects you can see. Another important advantage of a large telescope is its greater resolving power. With a bigger lens or mirror, you can see smaller details, like the surface markings on a planet, the binary companions of stars or the spiral structure in a distant nebula.

So size is everything, and astronomers have always been on a quest for larger instruments. But how big can a telescope be? Not so big if it's a refractor. Starlight has to pass *through* a lens so the lens can only be supported along its edge. If the lens is too large and too heavy it will sag. In contrast, the primary mirror of a reflecting telescope can be supported from the back. It can be much bigger. A mirror has another advantage, too. It only needs to have a perfect reflecting *surface*. A telescope lens, in contrast, has to be of perfect quality throughout. Every bubble or impurity in the glass will degrade the image.

In 1893 the largest refractor in history was presented at the World's Columbian Exposition in Chicago. Four years later it was installed at the University of Chicago's Yerkes Observatory in Williams Bay. Its lens, ground by famous optician Alvan Clark, is just over one metre across; its tube eighteen metres long. Astronomers have used this impressive instrument to study binary stars, stellar motions and distances and the spectroscopic properties of starlight, which have provided important clues to the chemical makeup of stars. But with the Yerkes Telescope, refractor builders had reached their limit. How to do better? Think mirrors.

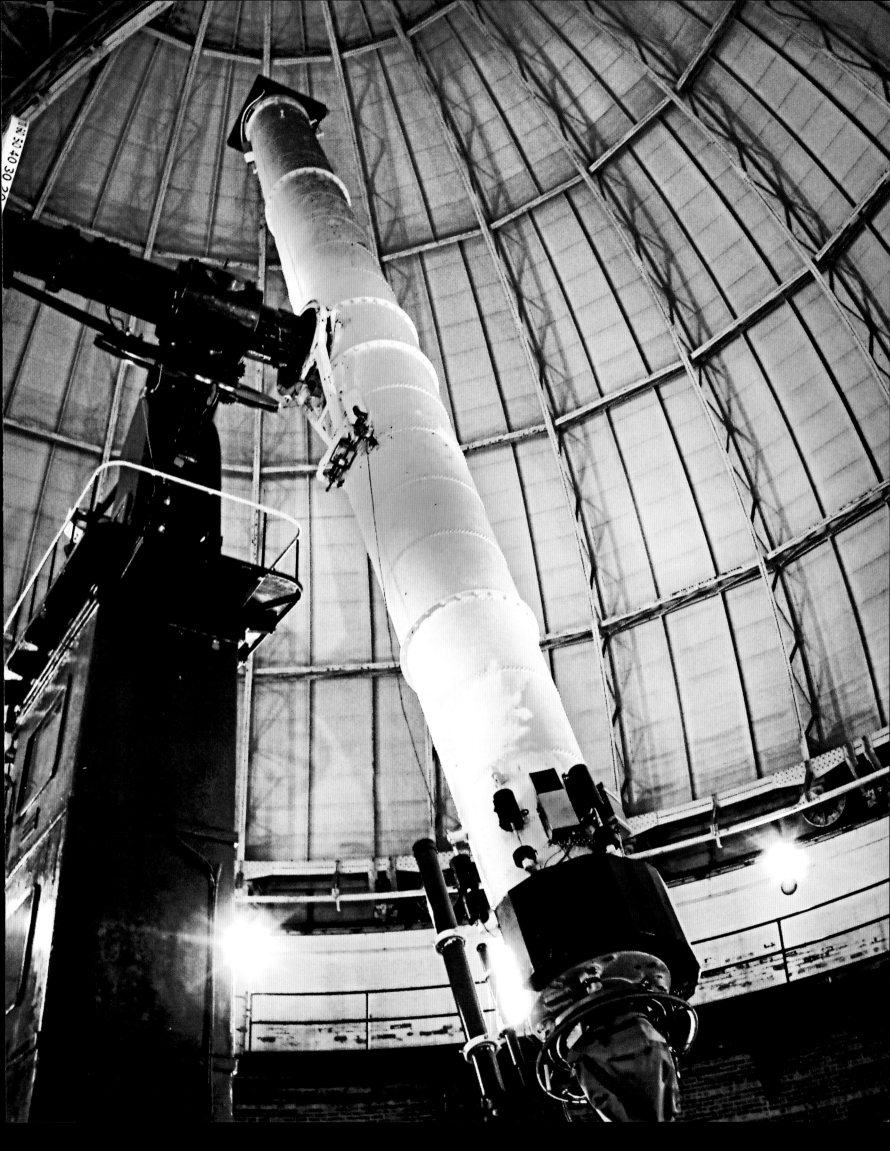

" An immense piece of cosmic artillery, ready to attack the Universe "

Summit seekers

For centuries astronomers carried out their observations wherever they happened to live, be it Padua, The Hague, Bath or Parsonstown. But by the second half of the 17th century, Isaac Newton had suggested that high mountaintops might be better suited for astronomical observations, because of the "serene and quiet air" that was to be expected there.

However, it wasn't until 1856 that Charles Piazzi Smyth — an Italian-born astronomer who served as Scotland's Astronomer Royal from 1846 to 1888 — proved Newton right. As part of his honeymoon, Smyth visited Tenerife in the Canary Islands and carried out test observations that convincingly revealed the merit of mountaintop observing.

Twenty years later the construction of the first permanently occupied mountaintop observatory commenced in California. East of San Jose, overlooking what is now known as Silicon Valley, Lick Observatory (named after philanthropist James Lick) was erected on Mount Hamilton, a 1284-metre high peak in the Diablo Range. Later, George Ellery Hale established the astronomical observatories at Mount Wilson (1740 metres) and Palomar Mountain (1713 metres). Kitt Peak Observatory (above, 2096 metres) was established in 1958.

Nowadays all major observatories are located at high altitude, where the air is not only cleaner and steadier, but also less dense. The best sites are peaks close to the ocean, where the airflow is particularly stable. Of course, a dry climate, cloudless skies and a sparsely populated area are other important factors. By far the largest high-altitude observatory is the Mauna Kea Observatory on the Big Island of Hawaii. Here, some of the largest telescopes in the world are located at 4200 metres above sea level.

The relatively small Mt. Evans Meyer-Womble Observatory of Denver University, at 4312 metres in the Rocky Mountains is even higher, as are the Indian Astronomical Observatory close to the village of Hanle, at 4517 metres in the western Himalayas, close to the Chinese border. And at 5000 metres above sea level in the Chilean Andes, on Llano de Chajnantor, Europe, the United States and Japan are constructing the Atacama Large Millimeter Array, which picks up millimetre waves from space before they are absorbed by atmospheric water vapour.

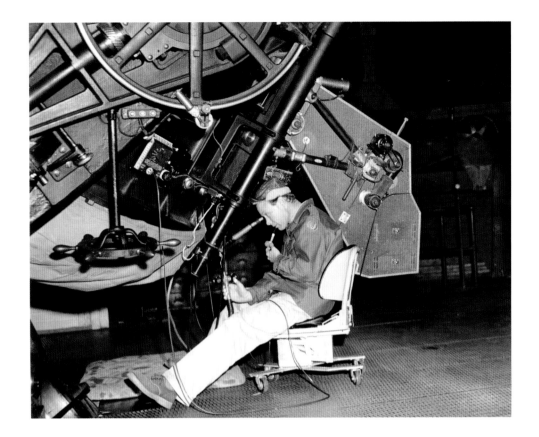

Big mirrors came to southern California a century ago, when Mount Wilson was a remote peak in the wilderness of the San Gabriel Mountains to the northeast of Los Angeles. Here, solar astronomer George Ellery Hale, who had also established the Yerkes Observatory, first built a 1.5-metre telescope. Smaller than Lord Rosse's Leviathan, but of much better quality. And at a much better site, too: the combination of the altitude of 1740 metres and the clear Californian skies yielded crisp images with unsurpassed detail.

Hale then talked local businessman John Hooker into financing a 2.5-metre instrument. Tons of glass and riveted steel were hauled up Mount Wilson. The Hooker Telescope was completed in November 1917 and would remain the largest telescope in the world for thirty years. An immense piece of cosmic artillery, ready to attack the Universe.

For the astronomers of the time, it must have been a wonderful experience to sit at the eyepiece of this giant instrument, sailing to other planets, floating through star clusters, and drifting past wispy nebulae. But the human eye was no longer the only device to catch the light from distant suns, starlight was also collected on photographic plates left in the focal plane of the Hooker Telescope for hours on end. Faint details and structures were revealed that no eye would ever have been able to see. Never before had scientists looked so deep into the cosmic void.

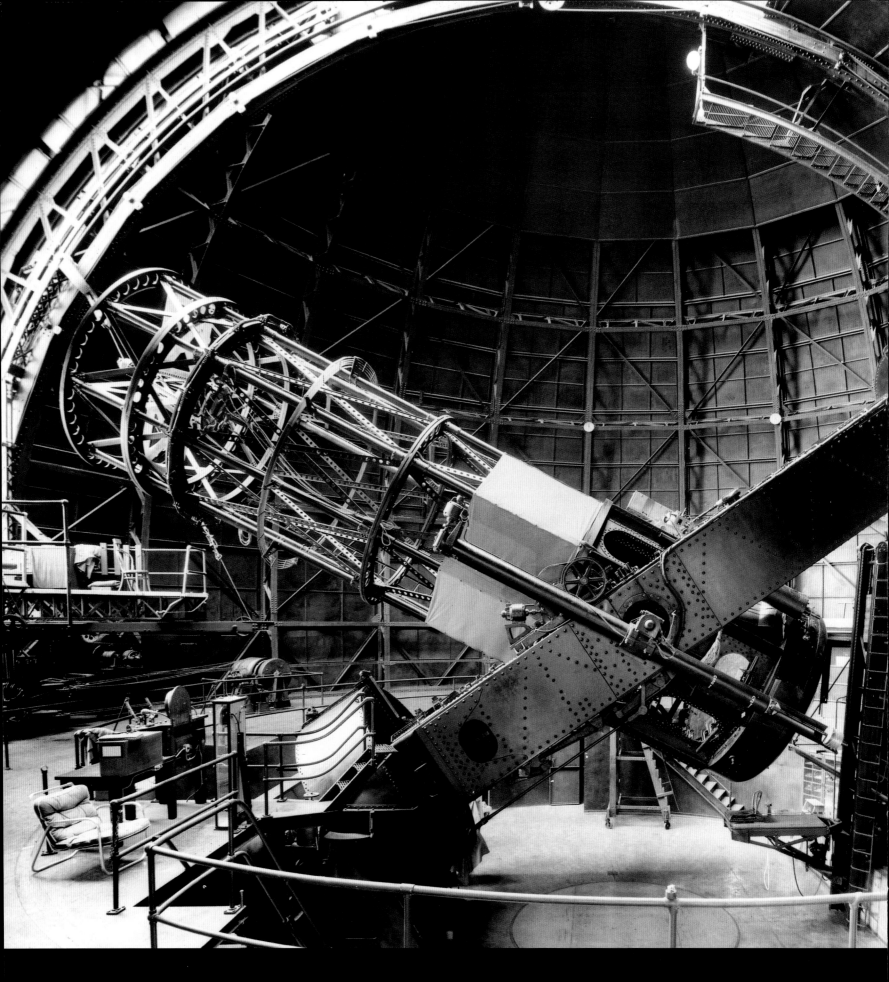

THE HOOKER TELESCOPE

This giant piece of cosmic artillery, the 100-inch (2.54-metre) Hooker Telescope at Mt. Wilson — named after the Los Angeles businessman who financed the project — not only revealed the true extragalactic nature of spiral nebulae, but also led to the discovery of the expansion of the Universe. In a sense, the Hooker Telescope gave us the Big Bang.

> *" The Hooker Telescope led scientists to one of the most profound discoveries of the twentieth century "*

Spiral nebulae appeared to be brimming with tiny stars. Were these nebulae really, as everyone supposed, whirlpools of gas in our Milky Way galaxy, or could they be sprawling stellar systems like our own? In April 1920 astronomers debated the issue fiercely, but there was no clear winner. Establishing the astronomical distance scale turned out to be an elusive endeavour. Photos taken with the Hooker Telescope resolved the mystery once and for all in 1923. Edwin Hubble discovered a star that changed brightness with clocklike precision in the Andromeda Nebula, one of the largest spirals in the sky. Such behaviour was well known from similar stars in our own Milky Way, whose true luminosities were known. This enabled Hubble to derive the star's distance: almost a million light-years away! Spiral nebulae were actually individual galaxies, well outside the realm of our Milky Way.

These spiral nebulae are receding from our own galaxy — something that can be measured with a spectrograph. Just as you hear a lower pitch when an ambulance siren is racing away from you, the observed wavelength of starlight is longer when its source is receding from the observer. Together with Milton Humason — a mule driver turned Mount Wilson janitor turned night assistant — Hubble studied this cosmic Doppler effect. They made a surprising discovery: nearby galaxies have small recession velocities, while more distant galaxies recede at a much higher pace. In 1927 Hubble concluded that the Universe must be expanding. The Hooker Telescope led scientists to one of the most profound discoveries of the twentieth century.

Father of big telescopes

George Ellery Hale (1868-1938) was a key figure in American astronomy. He was born in Chicago as the son of a wealthy elevator manufacturer, and inherited an interest in big construction projects. During his student years at the Massachusetts Institute of Technology he invented the spectroheliograph, used to study the light of the Sun. Although Hale was basically a solar astronomer, he is generally seen as the father of the big American telescopes. Through his efforts and fund-raising activities, he established Yerkes Observatory, Mount Wilson Observatory and Palomar Observatory. Hale also suffered from depression, which forced him to resign as director of Mt. Wilson in 1923.

" Life is a miracle in an eternally evolving Universe "

Thanks to the telescope we have traced the history of the cosmos. Our Universe hasn't always been there. It was born some fourteen billion years ago in an extremely dense and hot state known as the Big Bang — a huge explosive creation of time and space, matter and energy. Tiny quantum ripples grew into dense patches in the primordial brew, their gravity pulling ever more matter towards them. From these clumpy overdensities, whole galaxies condensed, in a stunning variety of sizes and shapes. Small galaxies collided and merged into larger ones, in which the formation of new stars spread like wildfire. Despite the expansion of space, gravity slowly drew galaxies together in compact clusters and filamentary superclusters, giving rise to the current large-scale structure of the Universe.

HALE TELESCOPE IMAGE OF THE ORION NEBULA

At some 1500 light-years from the Earth, the Orion Nebula is one of the nearest star-forming regions. Optical telescopes cannot penetrate the dense dust clouds to reveal the stellar cradles inside, but they are clearly visible in this near-infrared image, obtained with the Wide-field Infra-Red Camera (WIRC) at the 5-metre Hale Telescope on Palomar Mountain. New detectors like WIRC have given the venerable telescope a second lease of life.

Although the Universe started out as a simple mix of the two lightest elements, hydrogen and helium, nuclear fusion reactions in the cores of stars produced new atoms, including carbon, oxygen, iron and gold. Gentle stellar winds and devastating supernova explosions blew these heavy elements back into space as raw material for the formation of new stars and planets. And someday, somewhere, somehow, simple organic molecules evolved into living organisms. Life is a miracle in an eternally evolving Universe. The calcium in our bones, the iron in our blood and the oxygen we breathe were produced in the nuclear ovens of other suns. We are stardust.

THE NORTH AMERICA NEBULA

The North America Nebula (left) and the Pelican Nebula (right), named for their conspicuous shapes, are giant clouds of glowing gas in the constellation of Cygnus, the Swan. This image is a composite made from black and white plates taken with the Palomar Observatory's 48-inch (1.2 metre) Samuel Oschin Schmidt Telescope as part of the second National Geographic Palomar Observatory Sky Survey. The original images were recorded on two glass photographic plates — one sensitive to red light and the other to blue — which have since been digitised. A total of 62 individual frames were processed with the ESA/ESO/NASA Photoshop FITS Liberator by the Italian amateur astronomer Davide De Martin to create this colour view.

This broad-brush outline is the story of the Universe as it has been revealed to us over the past century through telescopic observations. Without the telescope we would only know about six planets, one moon and a few thousand stars. Astronomy would be stagnating, unable to move on from simple observations first made thousands of years ago.

"Like buried treasures, the outposts of the Universe have beckoned to the adventurous from immemorial times,"

wrote George Ellery Hale.

"Princes and potentates, political or industrial, equally with men of science, have felt the lure of the uncharted seas of space, and through their provision of instrumental means the sphere of exploration has rapidly widened."

THE FLAME NEBULA

Close to the star Alnitak — the left-most star in Orion's belt — is the star-forming region known as the Flame Nebula. Deep inside the dusty nebula is a newly formed cluster of baby stars, revealed in this dramatic near-infrared photo, which was obtained with the Wide-field Infra-Red Camera on the Hale Telescope. The Flame Nebula is located at some 1500 light-years from Earth.

> *" Hale had one final dream: to build a telescope twice as large as the previous record holder "*

As for instrumental means, Hale had one final dream: to build a telescope twice as large as the previous record holder. He secured a six million dollar grant from the Rockefeller Foundation, and selected a new site on Palomar Mountain, much further away from the increasing light pollution of Los Angeles. Thus emerged the grand old lady of twentieth-century astronomy — the 5-metre Hale Telescope, administered by the newly founded California Institute of Technology. Consisting of over five hundred tons of movable parts, it is so precisely balanced that it moves as gracefully as a ballerina. Its forty-ton mirror — cast from a new glass blend called Pyrex by the Corning Glass Works in New York, and transported to California by train — reveals stars forty million times fainter than the eye can see.

Completed in 1948 (ten years after Hale's death), the Hale Telescope has given us un-surpassed views of planets, star clusters, nebulae and galaxies. It has produced historic photos of giant Jupiter, with its many moons, and of the rings of Saturn. It has revealed the beauty of the Pleiades star cluster and faint wisps of gas in the Orion Nebula. And it has shown us famous galaxies like Andromeda, the Whirlpool and the Pinwheel as never seen before.

THE HALE TELESCOPE

A massive equatorial mount, with one rotational axis pointing toward the north celestial pole, supports the 200-inch (5.08-metre) Hale Telescope at Mount Palomar, which points straight up in this fish-eye photograph. The giant mirror is at the bottom of the telescope tube. It reflects starlight to the observer's cabin at the top. A secondary mirror can also reflect the light back through a central hole in the primary mirror to detectors at the rear of the instrument.

Another big refractor

The objective lens of the "Big Refractor" at Yerkes Observatory has a diameter of 40 inches — 101.6 centimetres. It's still the biggest refractor in history, but a new instrument on La Palma in the Canary Islands comes close. There, at 2400 metres above sea level on the rim of a volcanic caldera, the Swedish Institute for Solar Physics operates the Swedish 1-metre Solar Telescope (SST). Completed in 2002, it is currently the best solar telescope in the world. To fight air turbulence, the telescope tube is evacuated and the main lens is part of the vacuum seal. The SST regularly observes details on the sun as small as 70 kilometres.

But could astronomers do even better? In the late 1970s, Soviet astronomers tried. High
in the Caucasus Mountains, close to the town of Zelenchukskaya, they built their *Bolshoi
Teleskop Azimutalnyi* (Great Azimuthal Telescope) — sporting a primary mirror six metres
wide. But it never delivered the hoped-for results. It looked as if telescope builders had
reached the limits of technology. Dreams of even bigger instruments were shattered.

So had the history of the telescope come to a premature end? Of course not. We now
have eight- and ten-metre telescopes in operation, and even bigger ones are on the drawing board. The solution? New technologies.

Big money

Many big American telescopes have been made possible through generous grants from rich industri-
alists and philanthropists. Chicago businessman Charles Yerkes donated money for the observatory
that was named after him; John Hooker paid for the 2.5-metre telescope on Mount Wilson and the
Rockefeller Foundation enabled the construction of the 5-metre telescope on Palomar Mountain. Later,
oil entrepreneur William Keck funded the 10-metre Keck Telescopes on Mauna Kea, Hawaii; Microsoft
co-founder Paul Allen provided money for a huge array of radio telescopes, and former Intel chairman
Gordon Moore has sponsored design studies for the future Thirty Meter Telescope.

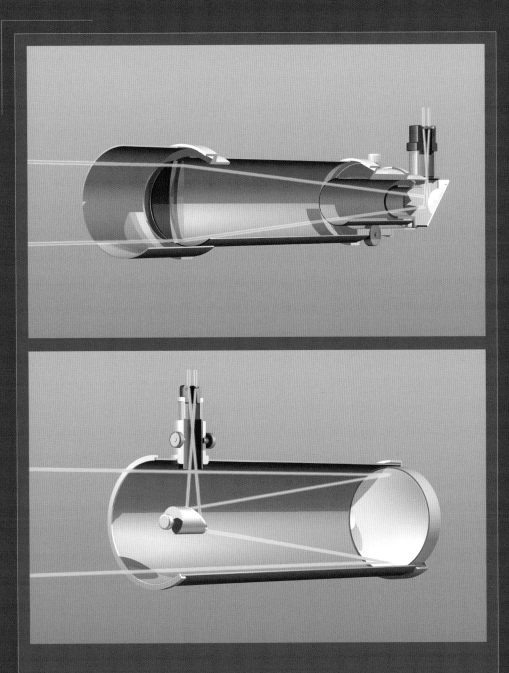

The two types of telescopes

The first telescopes used lenses — one objective lens and one or more lenses in the eyepiece — and were called refracting telescopes, or refractors. The refractor seen in the upper image is a modern design often used for amateur astronomy. Refractors were very popular up until the early 20th century, but became unmanageably large when the lens diameter went above 1 metre. Larger telescopes use mirrors — two or more — and are called reflectors. The lower image shows a Newtonian reflector design that is popular with amateurs who choose to build their own telescopes. Nearly all large, research-grade astronomical telescopes today are reflectors as they are easier to manufacture to a high standard and because it is easier to support a large mirror than a lens. A mirror can be supported from the backside and the effect of gravity can be counteracted as the telescope moves around to track the stars.

THE 3.5-METRE CALAR ALTO TELESCOPE

The 3.5-metre telescope seen here is housed at the Calar Alto Observatory located in the Sierra de los Filabres, Andalucía, Southern Spain. At an altitude of 2168 metres, the observatory was originally founded in the 1970s as a joint effort between Germany and Spain to provide astronomers from both countries with access to contemporary observational facilities in the northern hemisphere. Equatorially mounted 4-metre instruments like this one represented the state of the art of affordable telescopes in the 1970s.

EXTERIOR OF WIYN TELESCOPE

The WIYN Telescope, sporting a 3.5-metre mirror, is the second-largest instrument at the Kitt Peak National Observatory in Arizona, after the 4-metre Mayall Telescope. The instrument was completed in 1994 and is operated by a consortium of the University of Wisconsin, Indiana University, Yale University and the National Optical Astronomy Observatory (NOAO). Incorporating many new technologies, the telescope is one of the most powerful in its class.

TECHNOLOGY TO THE RESCUE

ESO'S NEW TECHNOLOGY TELESCOPE

The octagonal enclosure housing the European Southern Observatory's 3.6-metre New Technology Telescope (NTT) at Cerro La Silla in northern Chile was a technological breakthrough when completed in 1989. The telescope chamber is ventilated by a system of flaps that makes the air flow smoothly across the mirror, resulting in very sharp images. The NTT was also a testbed for fully computer-controlled alt-azimuth mounts, thin mirrors, and active optics.

Progress in telescopic astronomy would have come to a grinding halt in the second half of the twentieth century if it weren't for the digital revolution. Powerful computers have enabled a wealth of new technologies that have resulted in the construction of giant telescopes, perched on high mountaintops with monolithic or segmented mirrors as large as swimming pools. Astronomers have even devised clever ways of undoing the distorting effects of atmospheric turbulence and of combining individual telescope mirrors into virtual behemoths with unsurpassed eyesight. The optical wizardry of 21st century telescope building has ushered in a completely new era of ground-based astronomical discovery.

THE NEW TECHNOLOGY TELESCOPE PEERS INTO A STAR-FORMING REGION

A giant star-forming region known as the Omega Nebula reveals its dusty secrets to the near-infrared electronic eyes of ESO's New Technology Telescope. Located 5000 light-years away in the constellation of Sagittarius, the Archer, the Omega Nebula contains numerous young stars that are invisible to optical telescopes because of obscuring dust. At near-infrared wavelengths, however, most of the dust becomes transparent.

Just as modern cars don't look like Model T-Fords, current telescopes look very different from traditional instruments like the 5-metre Hale Telescope. Most obviously they have much smaller mounts. The mount is the support structure for the telescope tube. Astronomers want to be able to point a telescope wherever they like, so the mount consists of two perpendicular axes. By judiciously rotating the telescope about these two axes it can be trained on any point in the sky. But to keep an object in the eyepiece, the telescope also has to move continuously. The Earth's daily rotation means that not only the Sun, but the stars appear to rise in the east and set in the west. A telescope needs to track this apparent daily motion of the sky to keep the star under observation in the field of view.

Tracking the sky's motion becomes very easy if one of the two axes of the telescope mount points toward the Pole Star, the point on the sky about which the stars appear to rotate. To keep the stars in view, the telescope only has to rotate around this polar axis at the same constant speed as the stars. Known as an equatorial mount, this design was used in the big telescopes that were built in the first half of the twentieth century, but such mounts take up a great deal of space and are very heavy.

In contrast, the alt-azimuth mount, with one vertical and one horizontal axis, is much more compact. Here, the telescope is pointed like a cannon. Choose your bearing and elevation, and away you go. But tracking the motion of the sky becomes much harder. The telescope has to rotate at varying speeds around both axes at once, which needs precise computer control — something that only became available in the 1970s.

All the current big telescopes have computer-controlled alt-azimuth mounts. They are cheaper to build, and they fit into smaller domes — another cost-saving factor. Take the twin Keck Telescopes on Hawaii, for example. Their 10-metre mirrors collect four times more light than the Hale Telescope, but the Keck domes are smaller than the one on Palomar Mountain!

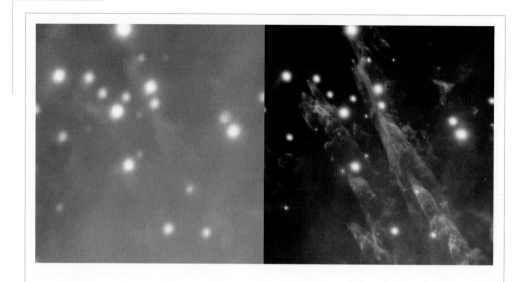

Untwinkling the stars

Adaptive optics sounds like magic. Starlight from the depths of space travels undisturbed for hundreds or thousands of years. But just before the light enters a ground-based telescope, the light waves are distorted by the Earth's lower atmosphere, resulting in twinkling stars and blurry images (left). As a result, even the biggest telescopes in the world can't see much more sharply than a medium-sized amateur instrument. So how can astronomers remove the blur and restore the view, as if there were no atmosphere?

The trick is to monitor a relatively bright star in the field of view and to measure the star's so-called wavefront. In space the wavefront is flat, but once in the Earth's atmosphere the wave has to cross air cells of different densities, so some wave crests reach the telescope a bit earlier than others, resulting in an undulating wavefront. If you know the actual shape of the wavefront by measuring light from a guide star, you can easily correct the image with a flexible mirror. Tiny heat-generating crystals on the back of this "rubber mirror" can adjust the mirror shape to compensate for the atmospheric distortions and restore the flat wavefront (right).

The rubber mirror is placed close to the focal plane of the telescope, where the starlight converges strongly and so the mirror can be relatively small — a few tens of centimetres or so across. Even then, adaptive optics faces a tremendous challenge, because the atmosphere is in constant motion. The whole cycle of measuring the guide star's wavefront, computing the necessary corrections and precisely adapting the shape of the flexible mirror has to be carried out about one hundred times per second!

Another disadvantage of adaptive optics is the need for a relatively bright guide star close to the observed object in the field of view. These natural guide stars are relatively rare, so astronomers are now using artificial guide stars that can be created at any point in the sky using a laser beam. The laser light excites sodium atoms in the upper atmosphere to generate a wavefront that can then be used to monitor atmospheric turbulence.

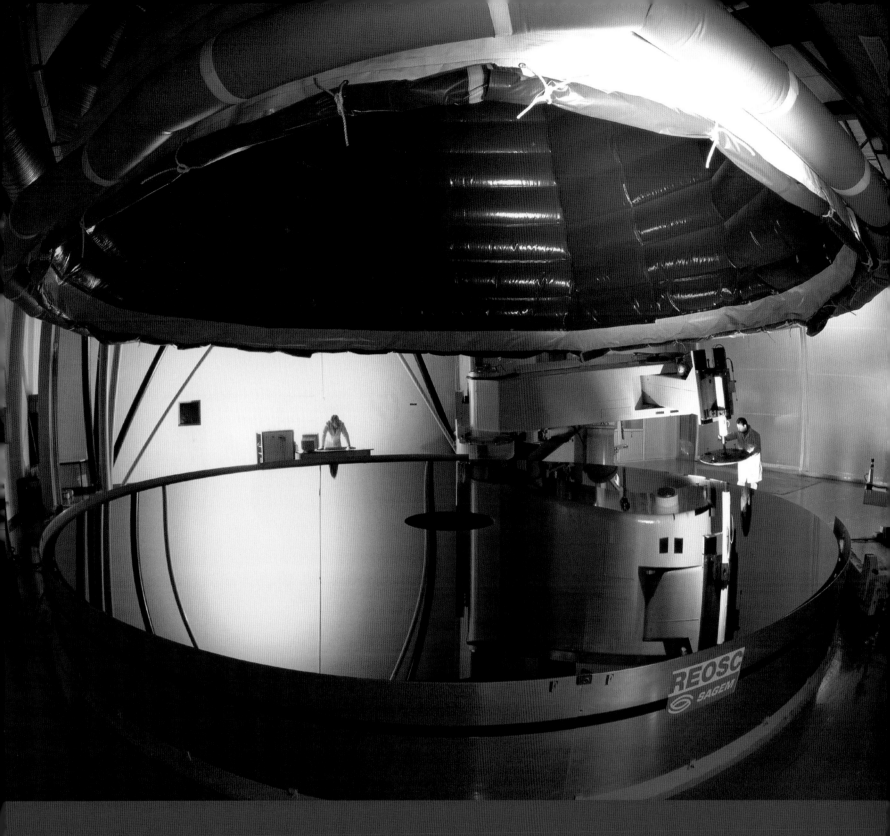

A VLT MIRROR

Smoother than a baby's skin, the fourth mirror of the European Southern Observatory's Very Large Telescope is shown here in December 1999, just after the final polishing phase was completed at the REOSC facility in Saint Pierre du Perray, just south of Paris. The meniscus-shaped mirror is 8.2 metres wide but only 20 centimetres thick.

> *" Optical engineers use giant, rotating ovens to cast meniscus-shaped mirror shells that can be many metres wide but less than twenty centimetres thick "*

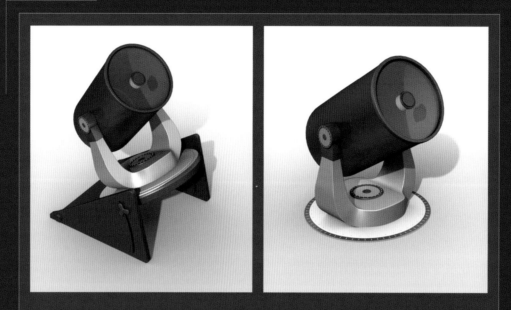

Telescope mounts

The development of new telescope mounts was one of the many technological advances for the telescope during the second half of the 20th century. The "old-fashioned" equatorial mount (left) moves about only one axis, parallel to the Earth's axis of rotation. This allows astronomers to follow the star they are viewing with ease. One simply has to maintain a slow but constant rotation of the telescope to keep it trained on a single patch of the night sky. Contrasting this, the alt-azimuthal mount (right) moves along two perpendicular axes, one horizontal and one vertical. While these mounts are much smaller and easier to make, they are also more difficult to use as one must simultaneously move the telescope in the vertical and horizontal directions with varying speeds to keep it trained on a single star. The alt-azimuthal mount only became popular after computers were introduced in the 1970s, but led to less massive instruments, smaller domes and thus better image quality.

" Cathedrals of science, devoted to observing the heavens "

Telescope mirrors, too, have evolved. Originally thick and heavy, they are now thin and lightweight. Optical engineers use giant, rotating ovens to cast meniscus-shaped mirror shells that can be many metres wide but less than twenty centimetres thick. An intricate support structure prevents the thin mirror from cracking under its own weight. Computer-controlled pistons and actuators also help to keep the mirror in perfect shape. This technique of continuously micro-adjusting the telescope to optimise performance is known as active optics. The idea is to compensate for slight deformations due to sagging, wind load, or temperature changes.

With a thin mirror, a telescope weighs less. The whole structure, including the mount, can be much slimmer and cheaper. European astronomers built the 3.6-metre New Technology Telescope on Cerro La Silla in northern Chile in the late 1980s as a test bed for these novel ideas in telescope construction. Even the enclosure had nothing in common with traditional telescope domes — it looked more like a giant silo — but the New Technology Telescope worked flawlessly. Astronomers were confident they could now break the six-metre barrier!

The Mauna Kea Observatory is at the highest point in the Pacific Ocean, at the summit of a dormant volcano 4200 metres above sea level. While tourists enjoy the sun and surf on the beaches of Hawaii's Big Island, astronomers face chilling temperatures and altitude sickness in their quest to unravel the mysteries of the Universe. The two Californian Keck Telescopes at Mauna Kea are among the largest in the world. The Keck mirrors are a staggering ten metres across and wafer-thin, but they are not single chunks of glass. They consist of thirty-six hexagonal segments, tiled like a bathroom floor, each controlled to nanometre precision. Cathedrals of science, devoted to observing the heavens.

sembles a giant metal spider. The huge mirror cell, outfitted with computer-controlled pistons, always keeps the thin mirror in perfect shape, independent of the pull of gravity, temperature changes and wind load.

At night, the twin Keck Telescopes on Mauna Kea collect photons from the depths of the cosmos — their mirrors larger than all earlier telescopes combined. What will be tonight's catch? A pair of colliding galaxies, billions of light-years away? A dying star, gasping its last breath into a colourful planetary nebula? Or an extrasolar planet that might harbour life?

By far the biggest astronomy machine ever built is the European Very Large Telescope (VLT), located on Cerro Paranal, a 2635-metre peak high in the Chilean Atacama Desert — the driest place on Earth. The VLT, operated by the European Southern Observatory (ESO), is really four telescopes in one, each sporting an 8.2-metre mirror. They are called Antu, Kueyen, Melipal and Yepun — native Mapuche names for the Sun, the Moon, the Southern Cross and Venus. The huge mirrors were cast in Germany, polished in France, shipped to Chile and then slowly transported across the desert.

Aluminising telescope mirrors

Most telescope mirrors are made of ceramic glass. But no matter how precisely a mirror blank is cast, ground and polished, it still needs a thin coating of reflective aluminium, and this coating needs regular maintenance. A telescope mirror has to be re-aluminised in a giant vacuum tank every few years or so because the reflectivity of the coating slowly degrades over time. At Cerro Paranal in Chile, home to the European Very Large Telescope, a special Mirror Maintenance Building has been constructed to carry out this delicate task. The job is easier for the Keck Telescopes on Mauna Kea because the 36 mirror segments can be processed individually.

" At Paranal, you are about as close to the Universe as
you can be without leaving Earth "

> *" At sunset, the giant telescope enclosures open up, starlight rains down on the VLT mirrors and new discoveries are made "*

At Paranal, you are about as close to the Universe as you can be without leaving Earth. The rock-strewn desert eerily resembles the surface of Mars. ESO's Residencia, where guest astronomers eat, sleep and relax, looks like a luxurious, futuristic spaceship. At night the star-spangled Milky Way arches overhead and the cosmos appears close enough to touch. At sunset, the giant telescope enclosures open up, starlight rains down on the VLT mirrors and new discoveries are made. Sometimes a laser pierces the night sky, creating an artificial star high up in the air. Delicate wavefront sensors measure how the star's image is distorted by atmospheric turbulence. Fast computers and flexible mirrors correct the view, in effect "untwinkling" the stars. Called adaptive optics, this is the magic trick of present-day astronomy, and it is used by all big telescopes. Without it, the Universe would look blurred. But *with* adaptive optics the images are razor-sharp.

Another sleight of optical wizardry, known as interferometry, combines the light waves from two big telescopes, with the wave crests and troughs precisely lined up — a technique first pioneered by radio astronomers. The result: the two telescope mirrors act as if they were part of one monster mirror as large as their mutual separation. Interferometry reveals details that would otherwise only be visible with a 100-metre telescope. The twin Keck Telescopes on Mauna Kea regularly team up as an interferometer. In the case of the VLT, with *four* individual elements, there are also small auxiliary telescopes that can be moved around to give even better performance.

SHOOTING A LASER AT THE GALACTIC CENTRE

An orange sodium laser is aimed at the core of the Milky Way to aid adaptive optics observations made with the European Southern Observatory's Very Large Telescope. The laser creates an artificial star high in the Earth's atmosphere. Astronomers continuously monitor the ever-changing distortion of the light of this laser guide star, which is caused by atmospheric turbulence, to correct for image blurring and to obtain the sharpest possible pictures.

Other major telescopes can be found all over the globe. On Cerro Las Campanas in Chile, American astronomers have constructed the twin Magellan Telescopes. Named after astronomers Walter Baade and Landon Clay, the identical instruments each have 6.5-metre mirrors that can be used separately, but will also work together as an interferometer. On Mauna Kea the Japanese Subaru Telescope ("Subaru" is the Japanese name for the Pleiades) is a high-performance instrument with an 8.3-metre mirror, fitted with sensitive spectrographs and cameras. The international 8.1-metre Gemini North Telescope is also on Mauna Kea. Together with its southern-hemisphere twin at Cerro Pachón in Chile, it allows astronomers to observe the whole sky in a standard fashion.

Light buckets

The Hobby-Eberly Telescope (HET) at Mount Fowlkes in Texas and the Southern African Large Telescope (SALT) in Sutherland, South Africa, have segmented mirrors measuring 11 metres across. These are special instruments, mainly used for spectroscopy, which have significant disadvantages if they are compared to conventional telescopes. They can only observe a small part of the sky, and don't use the whole mirror surface. As the standard image quality is relatively poor they are not equipped with sensitive cameras. What they are particularly good at is collecting and analysing the faint light from distant galaxies. Such "light buckets" are very different from normal telescopes, and so HET and SALT are not often listed among the largest telescopes in the world.

GEMINI NORTH TELESCOPE WITH LASER GUIDE STAR

With its silver-coated primary mirror and its powerful adaptive optics system, the 8.2-metre Gemini North Telescope on Mauna Kea is ready to unravel new cosmic mysteries using a yellow laser to create an artificial guide star in the Earth's atmosphere to monitor atmospheric turbulence.

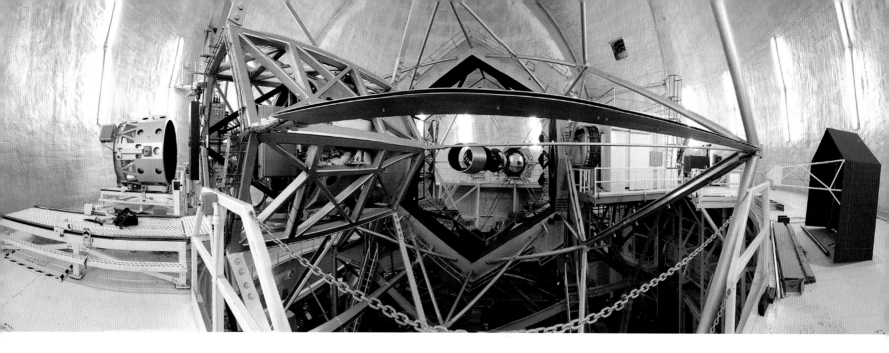

THE KECK TELESCOPE

It is impossible to cast
monolithic telescope mir-
rors larger than 8.4 metres
across using present-day
technology. Thus, the 10-
metre mirrors of the Keck
Telescopes at Mauna Kea,
Hawaii, consist of 36 smaller hexagonal segments — a
technique that will also
be used in the future for
extremely large telescopes.
The construction left of
centre in this extreme fish-
eye photo is the cage for the
secondary mirror.

" *These giant telescopes have eyes as large as swimming pools* "

The newest recruit is the revolutionary Large Binocular Telescope on Mount Graham in Arizona. Its two monolithic mirrors, both 8.4 metres wide, are mounted on the same support structure. Together, they have the light-collecting power of an 11.8-metre mirror. Using interferometry, they can see as much detail as a virtual telescope 22.8 metres across.

These giant telescopes are all constructed at the best available sites. High and dry; clear and dark. With eyes as large as swimming pools they are fitted with adaptive optics controls to counteract the blurring effects of the atmosphere. And occasionally they have the resolution of a virtual behemoth, thanks to interferometry.

The innovative technologies of giant, thin mirrors, adaptive optics and interferometry have provided us with exciting new views of the Universe. Astronomers have obtained ground-based images of the volcanoes on Io, the orange sulphur moon of distant Jupiter. They have measured the actual sizes — and squashed shapes — of some bright stars, like Altair and Regulus. They have even imaged a cool planet orbiting a brown dwarf, and tracked giant stars whirling around the very core of our Milky Way galaxy — governed by the huge gravitational pull of a supermassive black hole.

We've come quite a way since Galileo.

" We've come quite a way since Galileo "

**THE MCMATH-PIERCE
SOLAR TELESCOPE**

After a long day of observing the Sun, the McMath-Pierce Solar Telescope at Kitt Peak National Observatory is illuminated by twilight with the rising Moon in the background.

Remote control

Professional astronomers no longer peer through the eyepieces of their telescopes in the cold of the night. All major facilities are remotely operated from a comfortable control room where astronomers type commands on keyboards and peer at computer screens. In the case of the Keck Telescopes on Mauna Kea, the control room is at the Keck Observatory headquarters in Waimea, much closer to sea level, so guest astronomers don't need to suffer from altitude sickness. In many cases observatory staff astronomers and night assistants can carry out the observations, enabling most users to simply stay at home.

DUAL TELESCOPIC MIRRORS OF THE LARGE BINOCULAR TELESCOPE

Two eyes are better than one. The recently completed Large Binocular Telescope at Mount Graham International Observatory in Arizona sports two 8.4-metre mirrors on a single mount.

Together, they provide the light-gathering power of an 11.8-metre mirror and — through interferometry — the resolving power of an imaginary 22.8-metre telescope.

FROM SILVER TO SILICON

HUBBLE VIEW OF THE ORION NEBULA

Over five hundred Hubble Space Telescope images, in five different wavelength bands, have been stitched together to create this stunning mosaic of the Orion Nebula, one of the best-studied star-forming regions in the sky. The image shows over 3000 stars, most of which were born recently. Digital image processing makes it possible to combine optical and near-infrared data in one picture.

Observing the Universe through the eyepiece of a telescope is one thing, but recording the observations for posterity is something quite different. Originally astronomers used pen and paper to draw what they saw, but the human eye is a lousy detector and our brain can play tricks on us. Astrophotography, first explored in the mid-nineteenth century, has proved to be a powerful, objective way of recording telescopic images with the advantage that long exposures revealed much more than the eye could ever see. But the true revolution arrived with electronic detectors and digital image processing.

"For well over two hundred years, astronomers also had to be artists"

Four centuries ago, Galileo Galilei made pencil sketches of what he saw through his telescopes: the pockmarked face of the Moon, the dance of the jovian satellites, dark spots on the Sun and the stars of Orion. To share his discoveries he published his drawings in a small pamphlet called *The Starry Messenger*.

For well over two hundred years the astronomers who peered through their eyepieces and made detailed drawings of what they saw also had to be artists, sketching the stark landscape of the Moon, storms in the atmosphere of Jupiter or subtle veils of gas in a distant nebula. Although it was possible to use crosshairs, micrometers and precise timing devices to measure positions and dimensions on the sky accurately, the final depiction of the telescopic image remained a very personal, artistic endeavour.

Sometimes astronomers would draw features that didn't exist — like canals on Mars. Mars is a small and distant planet and the telescopes of the late nineteenth century were barely able to reveal surface details on its tiny, reddish disc. During rare, brief moments of perfect seeing, Italian astronomer Giovanni Schiaparelli picked out *canali* — dark linear features spanning the Martian globe. Percival Lowell in the United States saw them too. According to Lowell, these canals constituted a giant network irrigating the dry equatorial regions of Mars with water from the poles. Apparent signs of civilised life on the red planet!

It was all wishful thinking. There is no civilisation on Mars. We now know that the canals are an optical illusion, produced by the eye's tendency to find patterns everywhere. The human eye can be deceived: astronomers needed an objective technique for recording telescopic images.

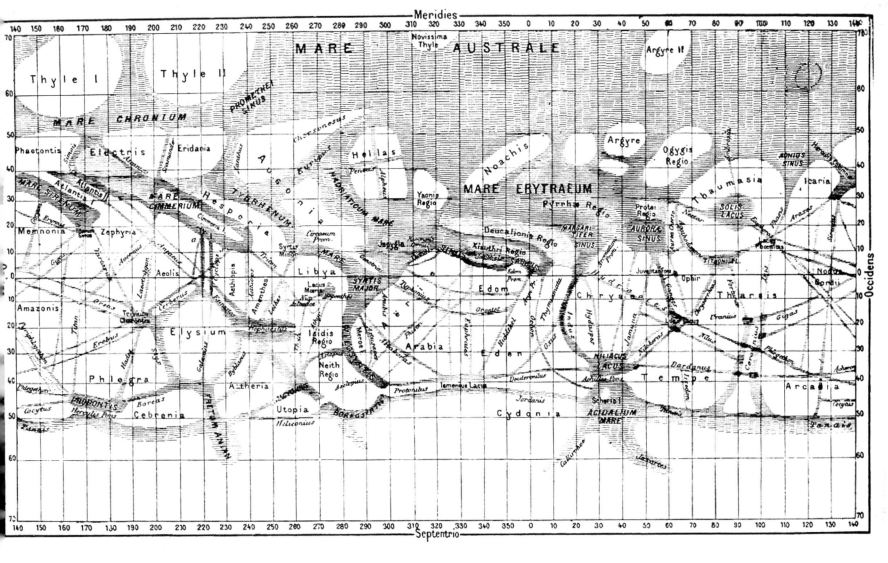

" Photography came to the rescue "

Photography came to the rescue. The first daguerreotype of the Moon was made by John William Draper in 1840. Photography was less than fifteen years old, but astronomers were already alive to its revolutionary possibilities. Ten years later, at Harvard College Observatory, astronomers took their first daguerreotype of a star, Vega. And in 1880, Draper's son Henry took the first photo of the Orion Nebula — no small feat, given how faintly the nebula glows.

GIOVANNI SCHIAPARELLI'S MAP OF MARS

The human eye can easily be misled. In the late 19th century, Italian astronomer Giovanni Schiaparelli thought he saw straight, dark lines on the surface of Mars. In this Mars map, based on his sketches and adorned with Schiaparelli's nomenclature, it is suggested that the Martian "canals" are indeed waterways connecting "seas" and dividing the terrain into a large number of different "lands". Astronomers later realised that the canals on Mars are an optical illusion.

THE FIRST PHOTOGRAPH OF THE MOON

On 18 December 1839, British-American scientist John Draper was the first person in history to take a photograph — or, to be more precise, a daguerreotype — of the Moon. Draper's moon shot was the start of astrophotography, a technical discipline that would completely revolutionise astronomy.

COLLAGE OF EARLY ASTROPHOTOGRAPHS

American optician George Willis Ritchey collaborated closely with George Ellery Hale and worked at the Mount Wilson Observatory for 14 years. Between 1901 and 1917, a period when astrophotography was still evolving, Ritchey obtained these photographic images of famous galaxies. Starting from the upper left and travelling clockwise, the images are of the Needle Galaxy (NGC 4565), the Triangulum Galaxy (M33), the Pinwheel Galaxy (M101), the Whirlpool Galaxy (M51), the central parts of the Andromeda Galaxy (M31) and M81 in Ursa Major.

Astrophotography is not only an objective way of observing the heavens, it can also reveal objects too faint for the human eye to see. The sensitive emulsion of a photographic plate contains small grains of silver halide. Expose them to light, and they darken. The result is a negative image of the sky, with black stars on a white background. And there is a bonus: a photographic plate can be exposed for hours on end, collecting more and more light, revealing ever fainter stars. Just gazing at the sky for longer won't make any more stars appear once the eye is dark-adapted. But photographs are different. A longer exposure reveals more stars.

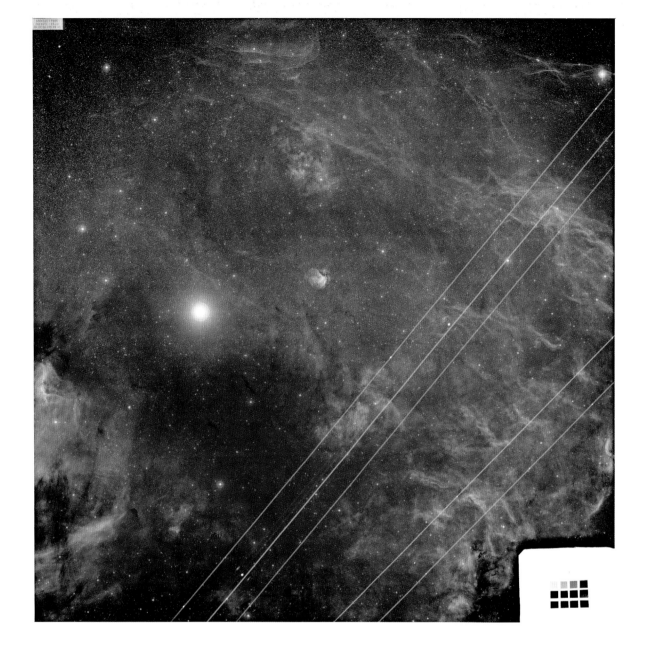

ASTROPHOTOGRAPHY WITH CHALLENGES

This full plate image from the Palomar Samuel Oschin Schmidt Telescope survey shows the possible defects that often plague astrophotographs. Passing diagonally across the lower right side are trails of light left by airplanes passing across the camera's field of vision during the long exposure. The vignetting of the telescope and camera is apparent as a darkening of the corners of the image. This is produced by a limitation of the optical design. The exposure sticker and a density sequence indicating the greyscales of the exposure are visible along the edge of the image. The brightest star in the image is Deneb in the constellation of Cygnus, the Swan, and the nebulosity on the left is part of the Pelican Nebula.

The Scottish astronomer David Gill, who worked at the Royal Observatory Cape of Good Hope in South Africa, was the first to realise the potential of astrophotography for charting the heavens. Earlier astronomers had compiled comprehensive star catalogues on the basis of meticulous visual observations — notably the 1862 *Bonner Durchmusterung*. Between 1895 and 1900 Gill photographed the sky systematically, revealing millions of stars. Dutch astronomer Jacobus Kapteyn then measured the plates, and together they published the three-volume *Cape Photographic Durchmusterung*, containing the positions and brightnesses of 454 877 stars.

" Astrophotography turned observational astronomy into a true science "

Photography became an indispensable tool for astronomers in the early twentieth century. Faint planetary satellites, the true nature of spiral nebulae, and the dwarf planet Pluto — all discovered using astrophotography. With the advent of big telescopes, sky surveys became bigger and deeper. The 1.2-metre Schmidt Telescope at Palomar Mountain was used to photograph the entire northern sky in the 1950s. The resulting Palomar Observatory Sky Survey, sponsored by the National Geographic Society, consisted of almost two thousand photographic plates, each exposed for nearly an hour and showing stars down to magnitude 21 — a million times fainter than the naked eye can see.

Astrophotography turned observational astronomy into a true science. Measurable, objective and reproducible. But silver is slow and exposure times are long, so to achieve results with photography astronomers have to be patient. The digital revolution has changed all that. Silicon has replaced silver. Pixels have replaced grains of halide. Even in consumer cameras, we no longer use photographic film. Images are now recorded on light-sensitive chips that were invented in 1969: charge coupled devices, or CCDs for short.

A CCD is a matrix of light-sensitive silicon elements — the pixels. During the exposure, each pixel accumulates an electric charge that is proportional to the amount of light it receives. After the exposure, the charges are electronically read out and converted into an image. The first two-dimensional CCDs measured 100×100 pixels or so, but megapixel cameras with five or six million elements are now found routinely in mobile phones.

The CCDs used by professional astronomers are extremely efficient. They are cooled using liquid nitrogen to well below freezing to make them even more sensitive so that almost every photon is registered. As a result CCD exposure times can be much shorter. A CCD mounted on a smaller telescope can now do in a few minutes what the Palomar Observatory Sky Survey achieved in an hour. Astronomers have already built huge CCD cameras with hundreds of millions of pixels and the silicon revolution is far from over.

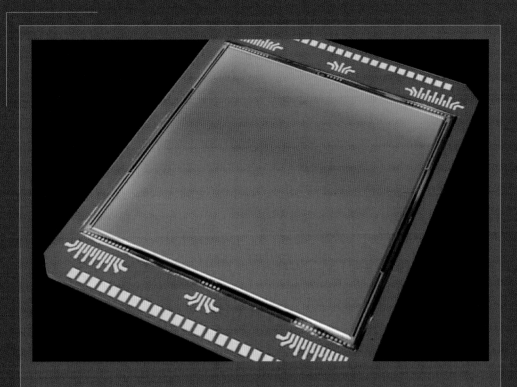

Cosmic colour

Thanks to the big telescopes, digital CCD detectors and the internet, the public at large is accustomed to stunning colour photographs of the sky: surrealistic hues in the atmosphere of Saturn, brilliant greens and reds in a star-forming region and a full rainbow of colours in planetary nebulae. Is the Universe really that flashy? Or do astronomers somehow fake the colours to make the pictures more appealing?

One thing is certain: an astronaut floating in front of the Eagle Nebula or the Crab supernova remnant would hardly notice the nebula, let alone see any colour, as it has an extremely low surface brightness. The human eye doesn't collect light like a photographic plate or a CCD detector, so the little light that the astronaut *would* notice would have no colour, since the colour-sensitive cones in the retina only respond to bright light. The "enhanced colour" often used in astronomical photographs highlights colour details we can never see for ourselves.

A modern CCD-detector is comparable to old-fashioned black and white film: it actually responds to a very wide spectrum of colours, but just records the total intensity. To create colour images from a CCD, colour filters are used to create a green, a red and a blue image, which can then be combined into a colour picture. This is how the well-known colourful Hubble images are created. With one important caveat: the filters are carefully chosen to provide as much scientific information as possible and do not necessarily correspond to the red, green and blue sensitive cells in our eyes. For instance, one filter is tuned to the wavelength of glowing oxygen gas, another filter to nitrogen or hydrogen. As a result, this type of final colour image can never correspond to the "true colours" our eyes would see.

Astronomers can also detect wavelengths that are invisible to the human eye, like ultraviolet and infrared radiation. In order to make this kind of information visible to our eyes, these wavelengths have to be represented with colours we can see. For instance, when a distant galaxy is observed through red, near-infrared and far-infrared filters, the resulting exposures might be allocated to the blue, green and red channels of a digital colour image.

The big advantage of digital images is that they are all set and ready for computer processing. Astronomers use sophisticated and specialised software to clean, tweak and
improve their snapshots of the sky. They can calculate results directly from the data in
the images and extract interesting information that can be plotted and visualised. The
images can be stretched or contrast enhanced to show the faintest regions of nebulae or
galaxies. Colour coding brings out structures that would otherwise remain invisible. By
combining three images made through different colour filters, glorious composites can
be produced, blurring the boundary between science and art.

There are other knock-on benefits from electronic astronomy. Never before has it been
so easy to find spectacular images of the cosmos. The newest Hubble photos or the latest
images from space probes orbiting distant planets are just one mouse click away! Semi-
professional amateurs use Photoshop to apply their image processing skills to images
taken with the big telescopes or to digitised versions of old photographic plates to create
strikingly beautiful colour composites for everyone to enjoy. Amateur astronomers with
small backyard telescopes that use cheap and simple webcam technology can routinely
obtain images of planets that show much more detail than the best photographs taken
with the 5-metre Hale Telescope half a century ago.

The digital revolution has also made it possible to automate the observing process completely. Robotic telescopes, equipped with sensitive electronic detectors, keep watch
over the sky. Advanced computer software compares images that are made a few hours
or days apart, picking up the slightest change to detect asteroids in the Solar System,
variable stars and extrasolar planets in our Milky Way or supernova explosions in distant
galaxies.

The art of blinking

For the better part of the twentieth century, astronomers have been using blink comparators in their
hunt for asteroids, variable stars and other changes in the sky. The idea is simple. Two photographic
plates of the same part of the sky, each containing tens of thousands of stars, are compared side by side
by looking through the eyepiece of the blink comparator. A mirror flips the observer's view between
the two plates every second or so. For most stars, the observer won't notice any difference, but if a
star changes position or brightness, it will appear to jump to and fro, or to switch on and off. With the
advent of digital images and powerful computers, this tiresome technique has finally become obsolete:
sky images can now be compared automatically simply by subtracting one from the other to reveal any
changes between the exposures.

" *Supernovae explode, new stars are born.*
Pulsars flash, gamma-ray bursts
detonate and black holes accrete "

" The LSST will open up a webcam window on the Universe "

Sky surveys have also become fully electronic and much more sensitive. The 2.5-metre Sloan Telescope in New Mexico has photographed and catalogued over a hundred million celestial objects, measured the distances to a million galaxies, and discovered a hundred thousand new quasars since the year 2000. Producing 200 gigabytes of data per night, the Sloan Digital Sky Survey has used thirty CCD chips, five colour filters and a novel scanning technique to map one quarter of the sky.

But one survey is not enough to monitor a Universe in constant flux. Comets come and go. Asteroids zip by. Distant planets orbit their mother stars, temporarily blocking part of the star's light. Supernovae explode, new stars are born. Pulsars flash, gamma-ray bursts detonate and black holes accrete.

To keep track of celestial events, astronomers want to carry out all-sky surveys every year. Or every month. Or twice a week. That's the ambitious goal of the Large Synoptic Survey Telescope (LSST), to be constructed at Cerro Pachón in Chile. This powerful instrument will have an 8.4-metre mirror, a field of view as large as fifty full Moons, and a three-gigapixel camera that continuously takes 15-second exposures of the sky. The LSST will open up a webcam window on the Universe when completed in 2015.

THE SLOAN 2.5-METRE TELESCOPE

The Sloan Digital Sky Survey 2.5-metre telescope at Apache Point Observatory, New Mexico, houses an incredibly complex digital camera, possibly the most complicated ever constructed. Inside are 30 charge coupled device (CCD) detectors cooled to minus 80 degrees Celsius. The Sloan Telescope has automatically mapped the northern sky in unprecedented detail.

Bad pixels

Before an electronic image of the night sky reaches the pages of an astronomy magazine or a coffee-table book, it needs to be thoroughly cleaned up. Different CCD pixels may respond differently to the same light levels, so the overall picture has to be precisely calibrated. Bright stars in the field of view may have caused a "bleeding" effect in a particular row of CCD pixels. And cosmic rays (energetic particles from deep space) will also increase the accumulated charge of a pixel. All these effects have to be taken care of before a colour image can be composed.

> " *A few years from now, anyone will be able to explore the cosmos from a laptop computer* "

The LSST is expected to produce thirty thousand gigabytes (or 30 terabytes) of data per night — a torrent of astronomical information that will be processed, analysed and managed in real time thanks to a promising partnership with Google. The results will be made available to the public so that a few years from now, anyone will be able to sail out into the Universe and explore the cosmos from a laptop at home.

Surveying the sky

The Large Synoptic Survey Telescope, due to become operational in 2015, will be by far the largest and most powerful survey instrument ever. But even before it is completed, astronomers will benefit from a number of other astronomical watchdogs, including Pan-STARRS on Hawaii — four 1.8-metre telescopes that will hunt for asteroids approaching Earth — and the 4.2-metre Discovery Channel Telescope of Lowell Observatory. Sponsored by Discovery Channel, this powerful instrument should become operational in 2010.

ARTIST'S IMPRESSION OF THE LARGE SYNOPTIC SURVEY TELESCOPE

Cerro Pachón in northern Chile has been selected as the location for the future Large Synoptic Survey Telescope (LSST), seen here in an artist's rendering. With its giant 8.4-metre mirror, large field of view, and huge digital camera, LSST will image the night sky once every three nights in a relentless search for short-lived phenomena and rapidly moving objects like supernova explosions and Earth-grazing asteroids. LSST could see first light in 2015.

5 | SEEING THE INVISIBLE

THE CONE NEBULA
The bright pink and red points in the centre of this stunning photograph are newborn stars that are only 100 000 years old. They are located in the Cone Nebula, a star-forming region in the constellation of Monoceros, the Unicorn. Thick layers of dust obscure these young stars in normal photographs, but they are clearly visible in this infrared Spitzer Space Telescope image.

The Universe is a black void, with a scattering of stars, nebulae and galaxies — or so it appears to observers using visible light. But if we include other forms of radiation invisible to us, the picture changes completely: clouds of interstellar hydrogen gas, emitting radio waves; stellar nurseries, glowing in the infrared; explosive outbursts of gamma rays and the all-sky background hiss of the Big Bang, diluted by almost fourteen billion years of cosmic expansion. So how do astronomers learn about the unseen Universe? By building telescopes and detectors that can see the invisible.

> *" Studying the Universe using visible light alone is like attending a concert with a severe hearing problem "*

Listen to your favourite music and you will hear the full spectrum of sound, from the deepest bass rumblings to the highest-pitched vibrations. But if your ears were sensitive to only a tiny part of the audible frequency range you would miss much of the performance. This was the problem for astronomers, confined to the visible by the limits of the human eye; they were locked into the middle register of the electromagnetic spectrum for hundreds of years.

Visible light consists of electromagnetic waves. Each colour corresponds to a certain wavelength. Red light has a wavelength of about 700 nanometres (0.0007 millimetres). Blue light is more energetic, with a higher frequency and a correspondingly shorter wavelength of about 400 nanometres. The human eye is sensitive to this optical range of colours, but not to electromagnetic waves with longer or shorter wavelengths. The Universe emits radiation at all wavelengths so studying the Universe using visible light alone is like attending a concert with a severe hearing problem.

The existence of most of the non-visible radiation arriving here from the depths of space was unknown until a century ago. Cosmic radio waves, for instance, were accidentally discovered in the 1930s. Although some of these waves have the same frequency as terrestrial radio stations, it doesn't mean that the Universe is broadcasting to us. Radio waves from space are extremely weak and there's nothing special to listen to — if you converted them into sound, you would only hear cracks and hisses. To "tune in" to the Universe, you need a radio telescope — usually a big dish. Radio telescopes can easily be much larger than optical telescopes as the dish surface doesn't need to be as smooth as an optical mirror since radio waves have longer wavelengths.

WESTERBORK SYNTHESIS RADIO TELESCOPE

Located in a rural area in the northeast Netherlands, the Westerbork Synthesis Radio Telescope is an array of fourteen 25-metre radio antennas lying in a straight line almost three kilometres long. Ten of the telescopes are permanently fixed along the east-west line while the other four can be moved along a railway system. The array was completed in 1970, but underwent a major renovation in 2000. On a regular basis, the Westerbork array is used in tandem with other radio telescopes from around the world in an observation mode known as very long baseline interferometry.

ANTENNAS OF CSIRO'S AUSTRALIA TELESCOPE COMPACT ARRAY

The Australia Telescope Compact Array, consisting of six 22-metre radio antennas in the Australian outback, is located some 500 kilometres northwest of Sydney. It is the premier radio interferometer in the southern hemisphere. The antennas can be moved around across an area six kilometres wide, to change between relatively wide-angle, low-resolution views and very detailed observations of a small part of the sky.

In the late 1930s, young American radio amateur, Grote Reber, built a 9.5-metre radio dish in his mother's backyard in Wheaton, Illinois. Reber made the first crude maps of the radio sky in which the Milky Way clearly stood out. Dutch astronomers Jan Oort and Henk van de Hulst immediately saw the enormous potential of the new technique: a radio telescope tuned to a wavelength of 21 centimetres could be used to map the distribution of cold, neutral hydrogen gas in the Universe. Very soon large radio telescopes were constructed all over the world, including the 76-metre Lovell Telescope in Jodrell Bank, England, which was completed in 1957.

VERY LARGE ARRAY ANTENNAS

Located just outside Socorro, New Mexico, the Very Large Array (VLA), in action since 1980, is a collection of 27 radio antennas, each measuring 25 metres across and weighing about 230 tons. Like other radio interferometers, the Y-shaped array functions as a single, giant telescope by electronically combining the data from all 27 antennas. Astronomers from around the world use the VLA to study everything from black holes to planetary nebulae.

It is also much easier to use interferometry at radio wavelengths and combine different telescopes to sharpen up images tremendously. The Very Large Array in New Mexico, for example, consists of 27 dish antennas, each measuring 25 metres across. Radio astronomers can combine individual telescopes that are spread out across continents. This technique, known as very long baseline interferometry (VLBI) has provided some of the most detailed observations in the history of astronomy.

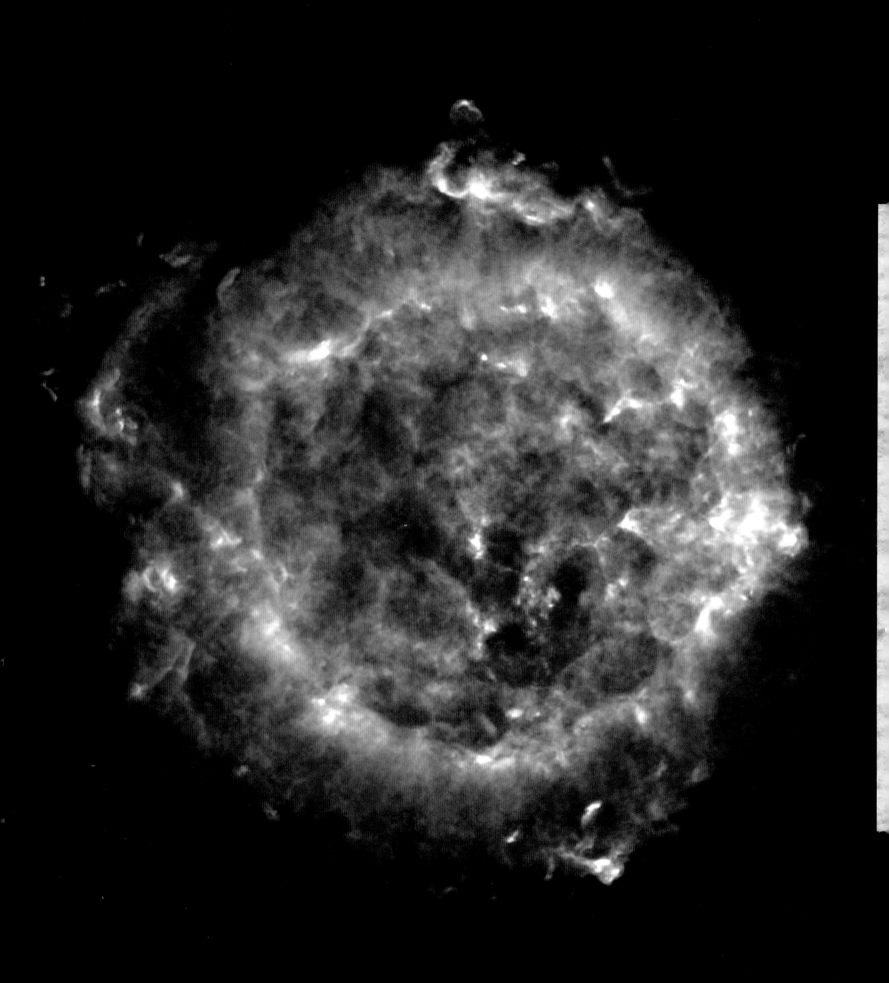

CASSIOPEIA A SEEN IN RADIO

Cassiopeia A is the remnant of a supernova explosion that occurred in our Milky Way over 300 years ago, at a distance of about 11 000 light-years. This radio image of Cassiopeia A, showing the ejected material as bright filaments, was created with the Very Large Array in New Mexico.

Dissecting light

There's more to electromagnetic radiation than just visible light. Radio waves, infrared (above) and ultraviolet radiation, X-rays and gamma rays are all variations on the same theme. Even visible light can be split up into its constituent wavelengths. White light, for instance, contains all the colours of the rainbow — something that becomes clear when sunlight falls on a piece of cut glass or on the tiny grooves in the surface of a DVD.

Using similar techniques (prisms and gratings), astronomers routinely dissect the light from stars, nebulae and galaxies into a so-called spectrum. A spectrum is a graph that shows the amount of energy received at various wavelengths. A hot, glowing object like the Sun emits light at every possible wavelength, but with a distribution that relates to the object's temperature. Extremely hot stars emit more blue light than yellow or red. Very cool stars emit most of their visible light at red wavelengths. For medium-temperature stars like our own Sun, the peak in the energy distribution lies right in the middle of the visible spectrum, which makes the Sun's light a roughly even mix of the colours, giving the Sun its whitish appearance.

Hot gases have a very different spectrum. The structure of the individual gas atoms determines the available energy levels so a hot gas only emits light at very specific, characteristic wavelengths that reveals its chemical composition to astronomers. For instance, hot sodium gas emits strongly in the orange part of the spectrum, as can be seen in low-pressure sodium street lighting. Cooler gases will also absorb light at these same characteristic wavelengths so the spectrum of a star can appear to be dimmer than expected at some wavelengths as gases in the star's atmosphere absorb light.

Spectroscopy — developed in the 19th century — has become an indispensable tool for astronomers. One important application is in stellar dynamics. The Doppler effect — an optical effect similar to the audible shift in pitch you hear when an ambulance siren moves towards or away from you — shifts the wavelengths of certain well-known spectral features by an amount determined by the motion of the emitting object. By measuring these wavelength shifts precisely astronomers can work out the radial velocities and rotation rates of stars and galaxies. Most significantly the expansion of the Universe would never have been discovered if it weren't for spectroscopy.

The main importance of spectroscopy, at all possible wavelengths in the electromagnetic spectrum, is its power to reveal the chemical composition of remote astronomical objects — something astronomers in the early 19th century could never have dreamed of.

"ALMA will be the largest — and highest — astronomical observatory ever built"

THE ALMA ARRAY

A few of the 66 antennas of the Atacama Large Millimeter Array (ALMA) can be seen in this artist's view of the observatory, which is under construction at Llano de Chajnantor in Chile, 5000 metres above sea level. The individual telescopes can be relocated in a variety of configurations by giant trucks to produce unsurpassed observations of the millimetre-wave emissions from remote galaxies or relatively nearby star-forming regions.

So what does the radio Universe look like? Our own Sun shines brightly at radio wavelengths and so does the core of our Milky Way galaxy. Clouds of hydrogen in the spiral arms of the Milky Way and in other galaxies glow at the telltale wavelength of 21 centimetres. Mysterious pulsars produce brief blips of radio waves. These are dense stellar corpses that spin around every few seconds, emitting rotating beams of radiation like a lighthouse. One bright radio source known as Cassiopeia A is actually the remnant of a supernova that exploded in the 17th century. Centaurus A, Cygnus A and Virgo A are giant galaxies that pour out huge amounts of radio waves, powered by massive black holes in their cores. Some of these radio galaxies and quasars are so energetic that their radio waves can be detected across ten billion light-years of empty space.

There's a faint, short-wavelength radio hiss that fills the Universe. Known as cosmic background radiation, it is the echo of the Big Bang; the afterglow of creation. The radiation was actually produced some 400 000 years after the birth of the Universe as energetic visible light, but it has now been diluted and stretched into the radio regime by cosmic expansion. Although the cosmic background radiation can best be mapped from space, in-depth studies have also been carried out with specially designed telescopes in Chile and Antarctica.

Each and every part of the electromagnetic spectrum has its own story to tell. At millimetre and sub-millimetre wavelengths, astronomers study the formation of the first galaxies in the early Universe and the origin of stars and planets in our own Milky Way. Most of this radiation is blocked by water vapour in the atmosphere so to observe it from Earth astronomers need to go high and dry.

Extreme astronomy

The Atacama Large Millimeter Array (ALMA), under construction in northern Chile, is the largest ever observatory project in ground-based astronomy and possibly the most demanding. Building and operating the 64-antenna array at five kilometres above sea level, where oxygen levels are half their normal value, is an unprecedented challenge. Site personnel will have a pressurised office area to prevent headaches and nausea, but the ALMA array will usually be operated remotely from a base camp at an altitude of "only" 3000 metres, outside San Pedro de Atacama.

To Llano de Chajnantor, for example. At five kilometres above sea level, this surreal plateau in northern Chile is the construction site of ALMA — the European/American/Japanese Atacama Large Millimeter Array. Here sixty-four antennas will work in unison as giant trucks move the 100-ton dishes around with millimetre precision. The dishes are either spread out over an area as large as London to see as much detail as possible or pushed close together to cover more area in the sky. When completed in 2014 ALMA will be the largest — and highest — astronomical observatory ever built.

> *" These instruments look right through cold cosmic dust clouds, revealing newborn stars that would otherwise be invisible "*

THE MAGIC TELESCOPE
At the Roque de los Muchachos Observatory on La Palma in the Canary Islands, night-time fog reveals the dozens of laser beams used to align the individual mirror segments of the 17-metre MAGIC Telescope precisely. MAGIC (Major Atmospheric Gamma Imaging Cherenkov) captures the extremely faint air-glow produced when energetic gamma rays from outer space create showers of elementary particles in the Earth's atmosphere. In 2008, an identical second telescope was completed, enabling stereoscopic observations.

The Universe also glows in the infrared. This "thermal radiation" was discovered by William Herschel and is emitted by objects at room temperature. On Earth, we can easily "see" in the infrared by using night vision goggles or cameras, but to study the weak infrared radiation from outer space, astronomers need extremely sensitive detectors, cooled to just above absolute zero.

Today, almost every big optical telescope has infrared cameras and spectrographs. These instruments look right through cold cosmic dust clouds, revealing newborn stars that would otherwise be invisible. Stellar nurseries, like those hidden deep inside the Orion Nebula and the Eagle Nebula, can be studied in exquisite detail at infrared wavelengths. Infrared detectors can also pick up the thermal radiation of circumstellar dust discs where new planets can form. Infrared observations are also good for studying very distant galaxies. Although their young stars shine brightly in the ultraviolet, this energetic radiation has travelled for billions of years through the expanding Universe before reaching Earth so its wavelength has been stretched out into the infrared regime.

And what about very high-energy electromagnetic waves, like ultraviolet and X-rays? Luckily for us, most of this harmful radiation is blocked by the Earth's atmosphere. The few ultraviolet rays from the Sun that reach the ground can cause skin cancer, but bursts of energetic X-rays could kill you. Cosmic gamma rays — the most energetic electromagnetic radiation in nature — produce cascades of high-energy particles when they hit the upper atmosphere. These particles, in turn, cause an extremely faint glow in the night sky that can be observed by instruments like the stylish MAGIC Telescope on La Palma.

PIERRE AUGER OBSERVATORY

Sixteen hundred jacuzzi-sized water tanks, spread out over an area of 3000 square kilometres, make up the Pierre Auger Observatory at Pampa Amarilla in Argentina. The water tanks, one of which is shown here, detect high-energy particles created when cosmic rays penetrate the Earth's atmosphere. Named after French cosmic-ray physicist Pierre Auger, the observatory was officially inaugurated in mid-November 2008.

" *Trillions of neutrinos travel through your body every second* "

Telescopes like MAGIC are not looking for any kind of radiation, but are particle collectors that pick up traces of energetic particles from the depths of space — electrons and atomic nuclei that have somehow been accelerated to nearly the speed of light. These cosmic rays, probably from distant supernovae and black holes, also create torrents of secondary particles when they collide with atomic nuclei in the atmosphere. The Pierre Auger Observatory in Argentina — named after the French physicist who pioneered cosmic-ray research — doesn't look like a telescope at all, but consists of 1600 detectors, spread over 3000 square kilometres.

And neutrino detectors — are they really telescopes? They do observe the Universe. Neutrinos are elusive particles that are produced in the Sun, in supernova explosions and in the Big Bang. Since neutrinos hardly interact with normal matter they can easily pass right through the Earth. Trillions of neutrinos travel through your body every second. So neutrino "telescopes" have to be extremely large to have a chance of picking up even one neutrino event. These detectors are built in deep mines, beneath the surface of the ocean or in the Antarctic ice to shield them from outside interference.

Cosmic ray tracing

Cosmic rays were first discovered by the Austrian physicist Victor Hess during high-altitude balloon flights in 1911, but the source of these energetic particles has been a mystery for many decades. Since their paths are bent by interstellar magnetic fields, the particles can't just be traced back to their point of origin. While many cosmic-ray particles are probably produced by shockwaves in supernova shells, the Pierre Auger Observatory has recently shown that the most energetic particles, each packing as much punch as a well-served tennis ball, originate in the immediate neighbourhood of supermassive black holes in the cores of distant galaxies.

" 'Telescopes' like LIGO are searching for tiny ripples in the very structure of spacetime "

Astronomers and physicists have also joined forces to build gravitational-wave detectors. Using laser interferometry, high-precision mirrors and evacuated tubes many kilometres in length, these "telescopes" are searching for tiny ripples in the very structure of space-time. Such gravitational waves are predicted by Albert Einstein's theory of relativity, but so far, none have been detected, although astronomers hope that may change when the Advanced Laser Interferometer Gravitational-Wave Observatory comes online in 2013.

Astronomers have opened up the full spectrum of electromagnetic radiation with a stunning variety of instruments and they have ventured outside the electromagnetic spectrum with cosmic-ray detectors and gravitational-wave observatories, but even with all this sophisticated kit some observations can't be done from the ground. And that's where space telescopes come in.

Pulsar proof

Gravitational waves — tiny ripples in spacetime, produced when large masses are strongly accelerated — have never been observed directly, but no one doubts that they exist. In 1974, American physicists Joe Taylor and Russell Hulse, using the 300-metre radio telescope at Arecibo, Puerto Rico, discovered the binary pulsar PSR 1913+16. The orbital period of the two pulsars is slowly decreasing, apparently as a result of the emission of gravitational waves: the energy loss exactly matches the predictions of Einstein's theory of general relativity. In 1993, Taylor and Hulse received the Nobel Prize in Physics for their discovery.

Telescopes observing the hidden Universe

The natural, bowl-shaped valley close to the harbour town of Arecibo, Puerto Rico, was transformed into the largest radio dish in the world and is one of the most remarkable telescopes ever built. The 300-metre Arecibo Observatory plays an important role in research on pulsars and in the Search for Extra-Terrestrial Intelligence (SETI). To observe even more exotic forms of light and radiation from the depths of the Universe, astronomers have constructed peculiar instruments that have little in common with Galileo's first telescopes. For these observatories, developed to study neutrinos. cosmic rays or gravitational waves, even the label "telescope" no longer seems to fit.

THE ATACAMA PATHFINDER EXPERIMENT TELESCOPE

The full Moon rises over the Atacama Pathfinder Experiment Telescope (APEX), located at 5000 metres altitude at Llano de Chajnantor in Chile. Using millimetre-wave observations from APEX, scientists hope to study warm and cold dust in star-forming regions both in our own Milky Way and in distant galaxies in the young Universe. They will be able to study everything from the structure and chemistry of planetary atmospheres to dying stars to molecular clouds as well as the inner regions of starburst galaxies.

THE JAMES CLERK MAXWELL TELESCOPE

A large membrane protects the dish of the James Clerk Maxwell Telescope (JCMT) from the harsh conditions at the summit of Mauna Kea on the Big Island of Hawaii. Sporting a 15-metre dish, the JCMT is the largest astronomical telescope in the world designed specifically to study sub-millimetre radiation from the cosmos. It is used to study many different things, including the Solar System, interstellar dust and gas, and distant galaxies.

6 | BEYOND EARTH

ASTRONAUTS AT WORK ON THE HUBBLE SPACE TELESCOPE

In December 1999, the Earth-orbiting Hubble Space Telescope was fitted with new gyroscopes, a new computer, and many other parts and instruments during a Space Shuttle Servicing Mission. Our home planet serves as a stunning backdrop for this wide-angle photo, which shows astronauts Steven Smith and John Grunsfeld at the tip of the Shuttle's robotic arm.

There's no better place for a telescope than space itself. Above the Earth's atmosphere observations are no longer hampered by air turbulence, so telescopic images of distant stars and galaxies are razor-sharp. Unlike a ground-based telescope, an instrument in Earth orbit can operate twenty-four hours a day and reach every part of the sky. Observing from space also makes it possible to study types of radiation that are otherwise absorbed by the atmosphere. Little wonder that the Hubble Space Telescope has made so many contributions to astronomy. And Hubble is not alone — more than 100 space observatories have been launched since the 1960s.

" Hubble has revolutionised every single field in astronomy "

The NASA/ESA Hubble Space Telescope is by far the most famous telescope in history. For good reason. It has revolutionised every field in astronomy. Hubble's mirror is small by current standards: only 2.4 metres across. But its location is — literally — out of this world. High above the blurring effects of the atmosphere, Hubble has the best possible view of the Universe. What's more: Hubble sees near-infrared and ultraviolet radiation that doesn't reach the ground. Cameras and spectrographs, some as large as a telephone booth, dissect and register the light from distant cosmic havens.

Like a telescope on the ground, Hubble can be upgraded. It was launched in April 1990 into a relatively low Earth orbit where it could easily be visited by NASA's Space Shuttle. Since then spacewalking astronauts have carried out Servicing Missions every few years. Broken parts have been fixed or replaced and older instruments have made way for new, state-of-the-art detectors. Hubble has become the workhorse of observational astronomy and has transformed our understanding of the cosmos.

Hubble has observed seasonal changes on Mars and a Saturn ring plane crossing, but the most spectacular event witnessed by the telescope was the impact of a comet on Jupiter in July 1994. The twenty fragments of comet Shoemaker-Levy 9 plunged into the atmosphere of the giant planet, producing huge fireballs and leaving giant dark markings that could easily be seen with an amateur telescope.

Looking beyond the Solar System, Hubble has followed the life cycle of stars from their birth and infancy in dust-laden clouds of gas to their final farewells as delicate planetary nebulae, slowly blown into space by dying stars or titanic supernova explosions that almost outshine their home galaxy. The famous "Pillars of Creation" in the Eagle Nebula have been shown to be the sites of future star formation. Deep in the Orion Nebula, Hubble has seen a breeding ground for new solar systems: dusty discs around newborn stars that may soon condense into planets.

HUBBLE ULTRA DEEP FIELD

Almost 10 000 remote galaxies light up in the Hubble Ultra Deep Field, the deepest visible-light image of the cosmos ever created. Cutting across billions of light-years, the snapshot includes galaxies of various ages, sizes, shapes, and colours. The smallest reddest galaxies seen in this picture may be among the most distant known, existing when the Universe was just 800 million years old.

Hubble's Deep Fields

On 9 March 2004 astronomers at NASA and ESA released the deepest-ever image of the distant Universe, showing no less than ten thousand galaxies out to distances of some thirteen billion light-years. This Hubble Ultra Deep Field (left) still serves as a rich cosmological goldmine, providing scientists with a unique way of studying the evolution of the Universe. But the Deep Field story started much earlier.

After the unfortunate near-sightedness of the Hubble Space Telescope's primary mirror had been corrected, Robert Williams, then director of the Space Telescope Science Institute, decided to use a large part of his director's discretionary observing time on a unique study of extremely remote galaxies. The idea was simple: take an extremely long exposure of one small, "empty" patch of sky, and see what comes out. Using Hubble's Wide Field and Planetary Camera 2 over a ten-day period in December 1995, 342 exposures of a small area in the constellation Ursa Major were made. The resulting image (above) revealed some three thousand distant galaxies, some of which turned out to be around twelve billion light-years away.

The strength of this first Hubble Deep Field was its use as a benchmark for follow-up studies with other telescopes, like the James Clerk Maxwell millimetre-telescope on Mauna Kea, Hawaii, the Very Large Array radio telescope in New Mexico, the European Infrared Space Observatory and NASA's Chandra X-ray Observatory. A similar approach was used in autumn 1998 to image the Hubble Deep Field South — a region of sky accessible to large optical and infrared telescopes in the southern hemisphere. In total, the two original Hubble Deep Fields have yielded hundreds of scientific papers, focusing on the early evolution and changing morphology of galaxies and on the star formation history of the Universe.

When the Hubble Space Telescope was fitted with the new, more sensitive Advanced Camera for Surveys in March 2002, it seemed only logical to carry out an improved version of the Deep Field study. A small area on the sky in the southern constellation of Fornax was imaged eight hundred times during autumn 2003, for a total exposure time of over eleven days. This Hubble Ultra Deep Field is still the best view of the distant Universe we have ever obtained.

"Without space telescopes, astronomers would be blind to energetic forms of radiation"

Near-sighted Hubble

Right after the launch of the Hubble Space Telescope, in the spring of 1990, horrified astronomers and technicians discovered that the Space Telescope's primary mirror had a minute deformation. The result: Hubble's images were not as sharp as planned. During the first Servicing Mission in December 1993, astronauts fitted Hubble with corrective optics to solve the problem and the Space Telescope has been performing beyond expectations ever since. The latest generation of cameras is designed to compensate for the spherical aberration of the primary, so the corrective optics are no longer necessary.

Hubble has studied thousands of individual stars in giant globular clusters — the oldest stellar families in the Universe. And galaxies. Astronomers have never seen so much detail. Majestic spirals, absorbing dust lanes, violent collisions. Extremely long exposures of blank regions of sky have revealed thousands of faint galaxies billions of light-years away by capturing photons that were emitted when the Universe was still in its infancy. These Hubble Deep Fields are astronomical windows into the distant past, shedding new light on the ever-evolving cosmos.

Hubble is not the only telescope in space — NASA's Spitzer Space Telescope, launched in August 2003, could be described as an infrared Hubble. Spitzer has an even smaller mirror than Hubble: just 85 centimetres across. The telescope itself is tucked away in a vacuum flask filled with liquid helium and its detectors, cooled to just a few degrees above absolute zero, are the most sensitive infrared detectors ever launched in space.

Spitzer has revealed a dust-filled Universe. Dark, opaque clouds of dust, invisible at optical wavelengths, glow in the infrared when heated from within. Shockwaves from galaxy collisions sweep up dust in spiral arms or in telltale rings announcing new sites of excessive star formation. Dust is also produced in the aftermath of a star's death. Spitzer has found that planetary nebulae and supernova remnants are laden with dust particles — the building blocks of future planets. Dust is even swirling in the strong winds of distant supermassive black holes. Using spectroscopy, Spitzer has been able to determine the chemical and mineralogical makeup of these cosmic dust particles.

SPITZER UNDER CONSTRUCTION AT LOCKHEED MARTIN

NASA's Spitzer Space Telescope, the most powerful infrared space observatory to date, receives a final check-up at Lockheed Martin Space Systems before its August 2003 launch. The 85-centimetre telescope is placed in a flask of liquid helium to keep it cool enough to detect the faint heat radiation from distant extrasolar planets, star-forming regions, and dusty galaxies.

101

"Energetic radiation passes right through a conventional telescope mirror"

At other infrared wavelengths Spitzer can look *through* the dust clouds and see the young stars hidden in their dark cores. Obscured star-forming regions, hidden from view by dark dust, become transparent. Spitzer's sensitive spectrographs have also studied the atmospheres of extrasolar planets — gas giants like Jupiter, racing around their parent stars in just a few days — to establish the existence of water vapour and sodium in these scorchingly hot atmospheres.

What about ultraviolet radiation, X-rays and gamma rays? They are completely blocked by the Earth's atmosphere. Without space telescopes, astronomers would be blind to these higher energy forms of radiation. Studying them is important to help understand the violent Universe of hot stars, supernova explosions, black holes, colliding galaxies and merging clusters. NASA's ultraviolet space telescope GALEX, launched in April 2003, has studied young, hot stars in hundreds of thousands of galaxies, showing astronomers how these cosmic building blocks evolve and change.

X-ray and gamma-ray telescopes are hard to build. Energetic radiation passes right through a conventional telescope mirror, while soft X-rays, with relatively low energy, can only be focused with nested mirror shells made of pure gold. Hard X-rays and gamma rays are studied with sophisticated pinhole cameras, or stacks of scintillators that produce brief flashes of light when hit by a high-energy photon. Despite these difficulties, astronomers have been lofting X-ray and gamma-ray instruments above the atmosphere since the early days of the space age. The first sources of cosmic X-rays were discovered by a Geiger counter on board a rocket probe in 1964.

Over the past decade, sophisticated high-energy space telescopes have been launched on board Earth-orbiting satellites. NASA's Compton Gamma Ray Observatory flew in the 1990s. At the time, it was the biggest and most massive scientific satellite ever launched — a full-fledged physics laboratory in Earth orbit. It made headlines with its detailed study of gamma-ray bursts — brief explosions of high-energy radiation, first detected by military surveillance satellites. In 2002 the European Space Agency launched its own gamma-ray observatory, INTEGRAL and in the spring of 2008, NASA deployed GLAST — the Gamma Ray Large Area Space Telescope.

Pinpointing gamma-ray bursts

NASA's Compton Gamma Ray Observatory confirmed that the mysterious cosmic gamma-ray bursts that occur on average once or twice per day are uniformly distributed across the sky. For many years astronomers weren't sure what this meant: the bursts could be relatively nearby phenomena in or around our own Milky Way galaxy or they could be titanic explosions in extremely remote galaxies. Astronomers using data from the Italian-Dutch satellite BeppoSAX observed the first optical afterglow of a gamma-ray burst in 1997 and found that the bursts do indeed occur billions of light-years away, making them the most powerful explosions in the Universe by far.

most energetic events in the Universe are the gamma-ray bursts: catastrophic terminal explosions of very massive, rapidly spinning stars. In less than a second, they release more energy than the Sun does in ten billion years.

Hubble, Spitzer, Chandra, XMM-Newton, Integral and GLAST are versatile giants. Other space telescopes are smaller and have more focused missions. Take COROT, for example — a French satellite devoted to stellar seismology and the search for extrasolar planets. Like NASA's future Kepler mission, COROT studies planetary transits: small, regular dips in the brightness of a star that occur when an orbiting planet is passing in front of the star's disc, as seen from the Earth.

THE SOMBRERO GALAXY OBSERVED WITH THE SPITZER SPACE TELESCOPE

The Spitzer Space Telescope and the Hubble Space Telescope joined forces to produce this striking image of one of the most popular sights in the Universe, the Sombrero Galaxy. It resembles a broad-brimmed Mexican hat when viewed in visible light, but the galaxy looks more like a "bull's eye" at infrared wavelengths.

HUBBLE AND CHANDRA MOSAIC OF THE CRAB NEBULA

The result of a supernova explosion in the year 1054, the famous Crab Nebula in the constellation of Taurus, the Bull, is seen here in data from the Hubble Space Telescope (green and dark blue), the Chandra X-ray Observatory (light blue) and the Spitzer Space Telescope (red). The bright white dot in the centre of the image is the corpse of an exploded star: a rapidly rotating and highly energetic neutron star with a mass equivalent to the Sun crammed into a sphere only twenty kilometres across.

" WMAP gave cosmologists their best view yet of the birth of the Universe "

Another small but powerful space telescope is NASA's Swift satellite — a combined gamma and X-ray observatory designed to unravel the mystery of gamma-ray bursts. Swift detects a few new gamma-ray bursts per week. It can determine the burst's position within a minute and radios the sky coordinates to robotic telescopes on the ground for follow-up studies at other wavelengths.

Then there is WMAP, the Wilkinson Microwave Anisotropy Probe. It has mapped the cosmic background radiation in unprecedented detail in just two years. Minute temperature fluctuations in this Big Bang afterglow provide information about density ripples in the early Universe that later grew into clusters and superclusters of galaxies. Combined with observations of the current large-scale structure of the Universe and of the expansion history of space — as revealed by the behaviour of distant supernovae — WMAP has given cosmologists their best view yet of the birth of the Universe, almost fourteen billion years ago.

Exploring the electromagnetic spectrum and opening up the space frontier are two of the most exciting developments in the history of the telescope. But the telescope is still young, and history isn't over. So what's next? Find out in the last chapter of this book, where we take a look into the future.

Big mirror

Late 2008 or early 2009 will see the launch of Herschel — a powerful infrared space telescope designed and built by the European Space Agency (ESA). Herschel, named after the scientist who discovered infrared radiation, has a primary mirror 3.5 metres across — the largest mirror ever built for a space telescope. Herschel will focus on far-infrared and sub-millimetre wavelengths, a part of the electromagnetic spectrum that hasn't been studied in detail yet. Its main purpose: to discover how the very first galaxies were born in the early Universe. Herschel will be launched together with ESA's Planck Surveyor, which is dedicated to studying the cosmic background radiation.

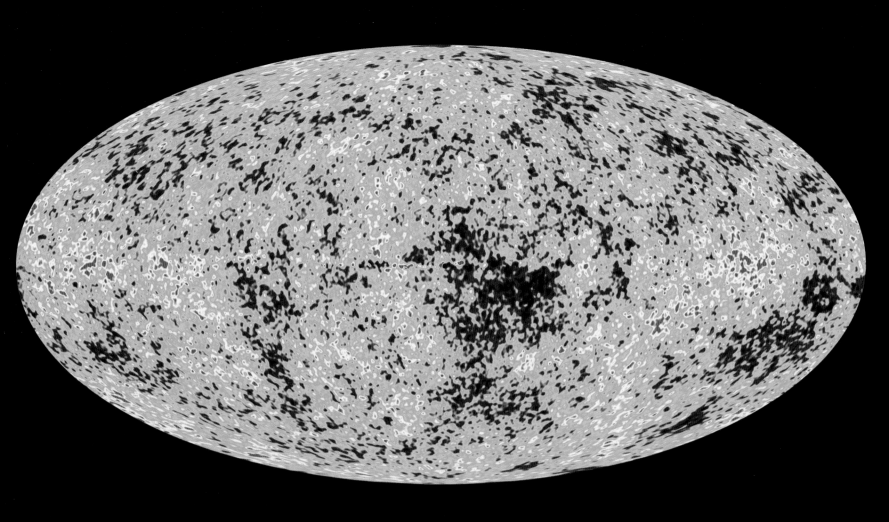

WMAP IMAGE OF THE MICROWAVE SKY

Minute temperature fluctuations in the very early Universe are represented by colour differences in this all-sky picture, which is based on three years of data from the Wilkinson Microwave Anisotropy Probe (WMAP) — a space telescope dedicated to studying the 13.7-billion-year-old cosmic microwave background radiation. The temperature fluctuations correspond to small density variations that grew into galaxies and clusters.

Space telescopes

Launching telescopes into space is expensive, but has allowed astronomers to peer into regions of the night sky they never dreamt would be accessible. These telescopes can view astronomical objects without the distorting effect of the Earth's atmosphere and have unhindered access to types of radiation we often can't observe from the Earth's surface: X-rays, infrared light, ultraviolet light and more. The observations from these telescopes sometimes provide almost unimaginable views of little known phenomena in the depths of space.

WHAT'S NEXT?

THE EUROPEAN EXTREMELY LARGE TELESCOPE

Spanning a monstrous 42 metres across, the primary mirror of the proposed European Extremely Large Telescope (E-ELT) consists of many hundreds of individual segments. When completed in 2017, the E-ELT will be by far the largest ground-based telescope for optical and near-infrared observations.

The telescope has been mankind's window on the Universe for four hundred years. It has provided scientists with unprecedented views of planets, stars and galaxies from our cosmic doorstep to the very depths of space and time. But despite their incredible performance, even the newest and most powerful telescopes leave room for improvement. Astronomers always want to venture beyond their current horizons. In this final chapter we take a look at things to come — the revolutionary ground-based telescopes and space observatories of the future. One thing is certain: there is much left to discover.

ARTIST'S RENDERING OF THE GIANT MAGELLAN TELESCOPE

Seven 8.4-metre mirrors, arranged like the petals of a flower, make up the Giant Magellan Telescope, which will be constructed

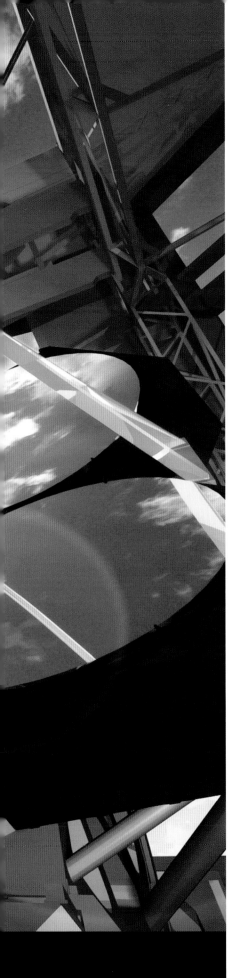

> *" Almost five hundred individual segments will make up one enormous mirror as tall as a seven-storey apartment "*

We've come a long way since the first telescopes of Hans Lipperhey, four centuries ago. And telescopic astronomy is far from finished. The best is yet to come.

The first mirror blank for the Giant Magellan Telescope (GMT) has already been cast at the Mirror Laboratory of the University of Arizona. This huge instrument will be built at the Las Campanas Observatory in Chile, which is already home to the twin 6.5-metre Magellan Telescopes. The GMT has no less than seven mirrors, arranged like the petals of a flower and each well over eight metres across. Together they will catch as much light as a 21.5-metre mirror and provide the same resolving power as a virtual 24.5-metre giant. Completion is expected in 2016.

The Californian Thirty Meter Telescope (TMT), due to be completed around the same time, is more like a giant version of Keck. Almost five hundred individual segments will make up one enormous mirror as tall as a seven-storey apartment. This will be able to collect ten times more light than the Keck Telescope and to see three times more detail. The 3.1-metre secondary mirror of the telescope — larger than the primary of the Hooker Telescope at Mount Wilson! — will be fully adaptive to compensate for atmospheric turbulence.

In Europe, plans are ready for the European Extremely Large Telescope (E-ELT) — an ambitious project led by the European Southern Observatory. This will also be a segmented-mirror telescope. But at 42 metres diameter, the E-ELT has twice the surface area of the American Thirty Meter Telescope. It has a revolutionary design that includes five mirrors and advanced adaptive optics to correct for the turbulent atmosphere. The E-ELT should become operational around 2017, probably somewhere in northern Chile.

CONCEPTUALISATION OF THE THIRTY-METER TELESCOPE

The impressive Thirty-Meter Telescope (TMT), currently under study by a consortium of Californian and Canadian institutes, will dwarf people and even freight trucks. This futuristic instrument will be built either on Mauna Kea, Hawaii, or on Cerro Armazones in the Chilean Atacama desert.

TMT

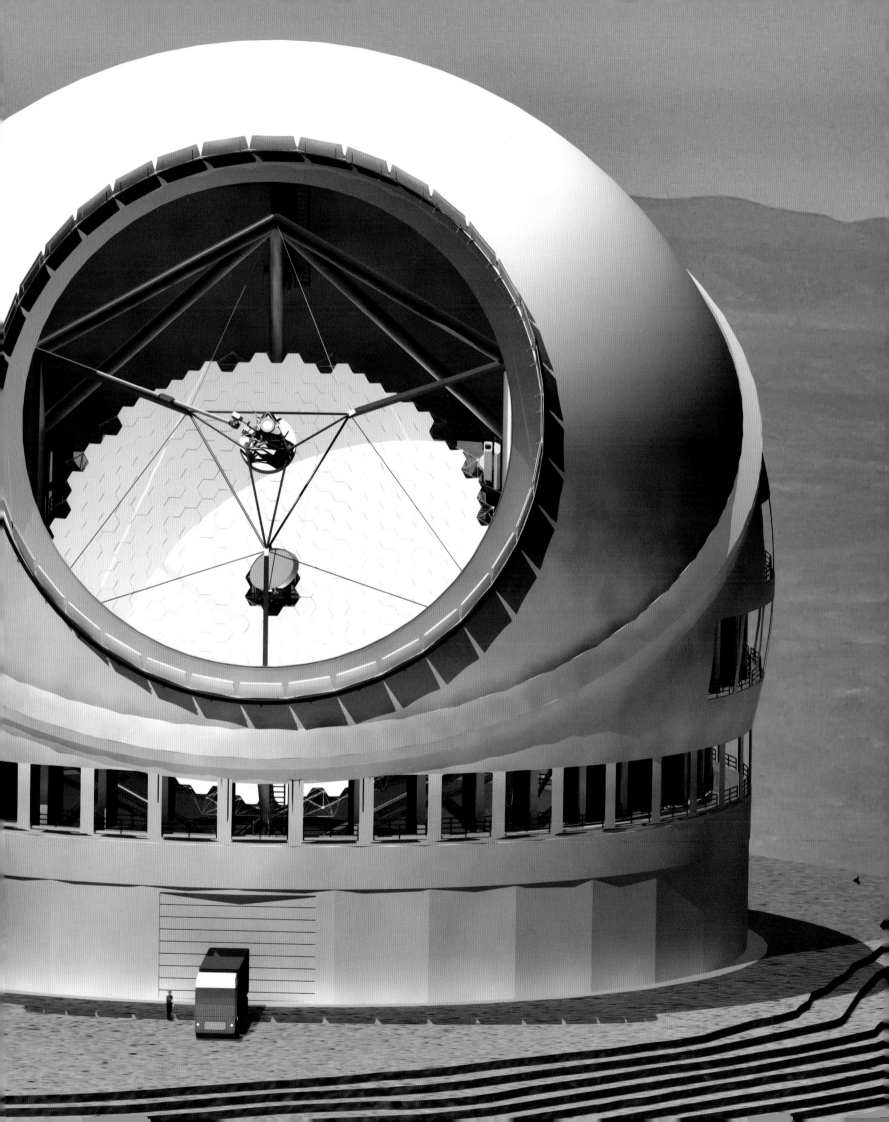

PANORAMIC VIEW OF THE EUROPEAN EXTREMELY LARGE TELESCOPE

The sun sets on the colossal dome of the future European Extremely Large Telescope. The E-ELT has a revolutionary design with five mirrors that involves advanced adaptive optics to correct for the atmospheric turbulence.

These future monster telescopes, optimised for infrared observations, will all have sensitive cameras, spectrographs and adaptive optics as standard. They will look back thirteen billion years to reveal the very first generation of galaxies and stars in the history of the Universe. They may also provide us with the first true picture of a Earth-like planet in another solar system — one of the Holy Grails of modern observational astronomy.

Awaiting E.T.'s call

The discovery of extraterrestrial life would be one of the biggest breakthroughs in history. However, it's unclear what the best search strategy might be. If life is pretty well ubiquitous, traces of fossil or extant micro-organisms might be found on Mars, right on our cosmic door-step. Looking somewhat further afield, biological activity would alter the chemical makeup of a planetary atmosphere, so future spectroscopic studies of extrasolar planets with space interferometers like the Terrestrial Planet Finder or Darwin might reveal the presence of alien organisms.

Unfortunately, these studies can only be carried out in our immediate galactic environment, at distances of a few tens of light-years at most. So if life is a rare phenomenon in the Universe, we may need to cast our net wider. That's where SETI comes in — the Search for Extra-Terrestrial Intelligence. The idea is that alien technological civilisations might betray their existence by beaming radio waves, or maybe even laser pulses, into space.

Ever since the early 1960s, radio astronomers have listened for artificial signals. The clockwork-precision radio blips of the first pulsar, discovered in 1967, were originally interpreted as an intelligent extraterrestrial message. But despite ever stronger efforts, no call has been received from E.T. It's unclear how this negative result should be interpreted; maybe looking for technological, radio-transmitting civilisations is just too anthropocentric.

A new generation of radio telescopes, including the Allen Telescope Array under construction in California and the future Square Kilometer Array, should be able to pick up signals from sources like ours on Earth as far away as the other side of the Milky Way galaxy. In a different part of the spectrum a relatively novel technique known as optical SETI is looking for brief, powerful flashes of laser light that might be used by alien civilisations to communicate across interstellar distances.

Whether we will ever establish contact with other life-forms in space is anybody's guess. Maybe complex, multicellular life is indeed extremely rare — which would be a milestone discovery in its own right. But the very first detection of extraterrestrial intelligence might be just around the corner. And using distributed-computing software like SETI@home, you can take part in the discovery.

" LOFAR will even look for possible radio signals from extraterrestrial civilisations "

DESIGN FOR THE SKA TELESCOPE

To be constructed either in Australia or in southern Africa, the international Square Kilometer Array (SKA) is a gigantic network of dish antennas and flat receivers that will provide astronomers with an unsurpassed view of the radio universe.

For radio astronomers, a mere 42 metres is peanuts. They can hook up many small instruments to synthesise a *much* larger receiver. The Low-Frequency Array, or LOFAR, is under construction in the Netherlands. Its thirty thousand inconspicuous antennas, grouped together in hundreds of "stations", are spread out over the country, with a few stations in neighbouring Germany. A dedicated network of fibre optics connects the antennas to a central supercomputer for data processing. This novel design has no moving parts, but it can still observe in eight different directions simultaneously. This may sound like magic, but is possible because the differences in the arrival times of radio waves from each location on the sky are known precisely for each and every antenna.

LOFAR technology will probably find its way into the Square Kilometer Array, which is now topping the wish-list of radio astronomers. This international array will be built in Australia or South Africa. Large traditional dish antennas and small stationary receivers will team up to provide incredibly detailed views of the radio sky. With a total collecting area of one square kilometre, the new array is going to be by far the most sensitive radio instrument ever built. Evolving galaxies, exploding quasars, blinking pulsars — no single source of radio waves will be safe from the spying eyes of the Square Kilometer Array. The instrument will even look for possible radio signals from extraterrestrial civilisations.

Atmospheric challenges

Without adaptive optics, building extremely large telescopes makes little sense. Adaptive optics (AO) uses wavefront sensors, fast computers, flexible mirrors and tiny actuators to compensate for the blurring effects of atmospheric turbulence. But current AO technology cannot easily be scaled up to the regime of 30- or 40-metre telescopes, because these giants catch starlight over such a wide area that the incoming wavefront has been affected by several different atmospheric cells at different altitudes. To solve this problem optical scientists are devising a technique known as multi-conjugate adaptive optics, using as many as five artificial guide stars created by sodium lasers on the ground.

> " *We can only speculate about the exciting discoveries Hubble's successor will make* "

THE JAMES WEBB TELESCOPE

From its distant vantage point, some 1.5 million kilometres behind the Earth, the 6.5-metre James Webb Space Telescope (JWST) is destined to be the successor of the Hubble Space Telescope. Unlike Hubble, however, the JWST will focus mainly on observations at infrared wavelengths.

And what about space? After its fifth and final servicing mission, provisionally scheduled for October 2008, the Hubble Space Telescope will be on active duty until 2013 or so. Around that time, its successor will be launched: the James Webb Space Telescope (JWST), a collaboration between the USA, Europe and Canada, named after a former NASA administrator. Once in space, its 6.5-metre segmented mirror will unfold like a flower and the JWST will be seven times more sensitive than Hubble. A large sunshade will keep the optics and the low-temperature instruments in permanent shadow.

The JWST won't orbit the Earth, but will be parked 1.5 million kilometres from our planet in a wide orbit around the Sun. Here it will have an unobstructed view of the whole sky and escape stray radiation from Earth. Half a century ago, the 5-metre Hale Telescope on Palomar Mountain was the largest in history. Soon, an even bigger telescope will be flying into the depths of space. We can only speculate about the exciting discoveries it will make. Stay tuned!

Other revolutionary space telescopes are being designed and planned. Some are relatively small optical instruments with one specific scientific goal, like the Kepler mission that will hunt for Earth-like extrasolar planets, or the SuperNova Acceleration Probe (SNAP), designed to study the expansion history of the Universe by observing distant supernova explosions. Others are medium-sized space telescopes like NASA's Wide-Field Infrared Survey Explorer and ESA's Gaia astrometry mission for mapping our Milky Way galaxy. And then there are the giant future space observatories like the European X-ray Evolving Universe Spectroscopy (XEUS) mission — a giant X-ray telescope consisting of two spacecraft flying in formation — and the international Laser Interferometer Space Antenna (LISA), which will hunt for gravitational waves.

"Radio astronomers want to put a LOFAR-like array of small antennas on the Moon"

" Mercury telescopes can only look straight up, but they're relatively cheap and easy to build "

Creative engineers come up with novel designs for new telescopes all the time. In Canada scientists have built liquid-mirror telescopes, in which starlight is reflected by the naturally curved surface of a rotating reservoir of mercury. These mercury telescopes can only look straight up, but they're relatively cheap and easy to build. They may turn out to be ideal instruments for statistical studies of remote galaxies, which can be found anywhere in the sky. There are even tentative plans for a Large-Aperture Mirror Array (LAMA) of eighteen 10-metre mercury telescopes working in unison, providing the same light-gathering power as the European Extremely Large Telescope.

Radio astronomers want to put a LOFAR-like array of small antennas on the Moon, way beyond terrestrial sources of radio interference. Someday there may even be a big optical telescope on the Moon's far side — supposedly the best possible place for optical astronomy in the inner Solar System. And using sensitive space telescopes and free-flying occulting discs, X-ray astronomers hope to improve their eyesight tremendously. They may even succeed in imaging the edge of a black hole.

LOFAR-LIKE RADIO ANTENNAE ARRAY ON THE MOON

In the distant future, astronomers may construct telescopes on the surface of the Moon. This artist's conception shows an array of radio antennas similar to the Low Frequency Array (LOFAR) that is under construction in the Netherlands. On the far side of the Moon, radio interference from terrestrial sources would be absent.

Mapping dark matter

Using a phenomenon known as weak lensing, astronomers are able to map the distribution of the invisible dark matter in the Universe. Strong gravitational lensing is the well-known effect by which the image of a background galaxy is distorted by the gravity of a foreground cluster. Weak gravitational lensing is much more subtle. Huge concentrations of dark matter will affect the image shapes of very remote galaxies ever so slightly. By studying the observed shapes of thousands or even millions of distant galaxies statistically, the intervening dark matter clouds (seen in blue above) can be mapped in some detail. This is one of the goals of the SuperNova Acceleration Probe (SNAP).

> *" It's hard to imagine that Earth is the only living planet in the entire Universe "*

One day the telescope may answer one of our most profound questions: are we alone in the Universe? We know there are other solar systems out there. We suspect there are extrasolar planets like Earth with liquid water. But... is there life? It's hard to imagine that our home planet is the only one in the entire Universe where the interaction of water and organic molecules — both ubiquitous in interstellar space — has led to the formation of living cells. But we don't know for sure. Future space interferometers may provide the answer. NASA has been studying a project called the Terrestrial Planet Finder and in Europe, scientists have proposed the Darwin array. Both should be able to sniff out extraterrestrial life.

Darwin's design consists of three or four giant space telescopes orbiting the Sun in formation, their mutual separations controlled by lasers to the nearest nanometre. Together, the telescopes would have enough resolving power to actually *see* Earth-like planets around other stars. The next step is then to study the small amount of light reflected by the exo-Earth that carries the spectroscopic fingerprint of the planet's atmosphere. In fifteen years time we may be able to detect the signatures of oxygen, methane and ozone — regarded by astrobiologists as the signposts of life — on a faraway planet. Aliens parking flying saucers in Time Square might be more spectacular, but finding micro-organisms elsewhere in space would finally prove that we're not alone in the Universe and could well be the biggest and most thought-provoking discovery ever.

ONE OF THE DARWIN MISSION TELESCOPES

Three or four space telescopes like the one shown in this image could work together as a giant space interferometer, like the Darwin array proposed by the European Space Agency. Darwin should be able to image terrestrial planets orbiting nearby Sun-like stars directly.

GRAND SPIRAL NGC 1672

Blue clusters of hot, young stars and pinkish clouds of glowing hydrogen gas trace the grand spiral arms of the galaxy NGC 1672, which is some sixty million light-years away in the southern constellation Dorado, the Goldfish. Delicate curtains of dust partially obscure and redden the light of the stars behind them. Four hundred years after the invention of the telescope, this Hubble Space Telescope image represents current state-of-the-art telescopic astronomy.

> *" Hundreds of thousands of amateur astronomers, all across the globe, go out every clear night to marvel at the cosmos "*

Amateur eyes

Dedicated amateur astronomers can make valuable contributions to professional astronomy. Not because they have bigger instruments, but because they have more time, collectively. Large professional telescopes are few, have to be booked well in advance and can only look at one particular thing at a time. But medium-sized amateur telescopes are numerous, and, thanks to computer control and electronic detectors, they can significantly add to the study of transient phenomena like stellar occultations by asteroids, transiting extrasolar planets, and gamma-ray burst afterglows. In some cases, amateur astronomers have even contributed enough to be listed as co-authors on a scientific paper.

The Universe is full of surprises. The sky never ceases to impress. Little wonder that hundreds of thousands of amateur astronomers, all across the globe, go out every clear night to marvel at the cosmos. They look at the Moon and the planets, they observe variable stars and comets, and they gaze at distant nebulae and galaxies. Their telescopes are much better than the simple instruments used by Galileo four hundred years ago. And thanks to the wonders of technology, their electronic images surpass the best photos made even a few decades ago by professionals.

In the quest for cosmic understanding, telescopic exploration of the Universe is just four centuries old. There is still a lot of uncharted territory out there. You can join the discoverers. Look up and wonder. Eyes on the skies.

Govert Schilling

Govert is a self-taught astronomy writer and populariser in the Netherlands. As a teenager, he started out as an amateur astronomer after seeing the planet Saturn through a three-inch telescope. Although Govert never lost his interest in astronomy, he did not pursue an academic career. Instead, he trained as a mechanical engineer and subsequently became a picture researcher for a Dutch encyclopedia publisher and editor-in-chief of the Dutch amateur astronomy monthly magazine *Zenit*.

In 1982 Govert was hired as a scriptwriter and programme editor by the planetarium in Amsterdam. He also began to write newspaper and magazine stories about astronomy, and his first book was published in late 1985. In 1987 the planetarium moved to Artis (the Amsterdam zoo) and Govert continued to produce popular planetarium shows, including a children's show based on Sesame Street characters, for which he received an Award of Excellency from Children's Television Workshop.

For many years Govert combined his part-time job for the Artis Planetarium with his freelance work for newspapers, magazines, radio and television. In 1998 he became a full-time freelancer. He writes about astronomy and space science for the Dutch daily national newspaper *de Volkskrant*, for a number of other Dutch weekly and monthly magazines, for *New Scientist* and *BBC Sky at Night* in the United Kingdom, and for *Sky & Telescope* (for which he is a contributing editor), *Science* and *Scientific American*.

Govert has written almost fifty books on various astronomical topics, from children's books and simple sky guides to topical books about new developments in astronomy. Some of his books have been translated into German and English, including *Flash! The Hunt for the Biggest Explosions in the Universe*, *Evolving Cosmos*, and *The Hunt for Planet X*.

He is also the owner and editor of the popular Dutch astronomy portal site www.allesoversterrenkunde.nl. In 2002 he received the prestigious Dutch Eureka Prize for his contribution to the popularisation of science and technology.

Asteroid 10986 was named Govert by the International Astronomical Union in 2007.

Lars Lindberg Christensen

Lars is a science communication specialist heading the Hubble European Space Agency Information Centre group in Munich, Germany, where he is responsible for public outreach and education for the NASA/ESA *Hubble Space Telescope* in Europe. He obtained his Master's Degree in physics and astronomy from the University of Copenhagen, Denmark. Before assuming his current position, he spent a decade working as a science communicator and technical specialist for the Tycho Brahe Planetarium in Copenhagen.

Lars has more than 100 publications to his credit, most of them in popular science communication and its theory. His other productive interests cover several major areas of communication, including graphical, written, technical and scientific communication. He has written a number of books, notably *The Hands-On Guide for Science Communicators* and *Hubble — 15 Years of Discovery*. His books have been translated to Finnish, Portuguese, Danish, German and Chinese.

He has produced material for a multitude of different media from star shows, laser shows and slide shows, to web, print, television and radio. His methodology is focussed on devising and implementing innovative strategies for the production of efficient science communication and educational material. This work involves collaborations with highly skilled graphics professionals and technicians. Some of the products of these collaborations are visible at: http://www.spacetelescope.org.

Lars is Press Officer for the International Astronomical Union (IAU), a founding member and secretary of the IAU Commission 55 Communicating Astronomy with the Public (http://www.communicatingastronomy.org), manager of the ESA/ESO/NASA Photoshop FITS Liberator project, executive editor of the peer-reviewed *Communicating Astronomy with the Public journal*, director of the *Hubblecast* video podcast, manager of the IAU International Year of Astronomy Secretariat and the Executive producer and director of the science documentary *Hubble — 15 Years of Discovery*. In 2005 Lars was the youngest recipient so far of the Tycho Brahe Medal for his achievements in science communication.

Image credits

p. 0-1
Subaru Telescope, National Astronomical Observatory of Japan

p. 4-5
Gemini Observatory

p. 6
Hubble Space Telescope, Chandra X-Ray Observatory and Spitzer
Space Telescope/NASA, ESA, CXC and JPL-Caltech

p. 8
ESA/Hubble (M. Kornmesser)

p. 11
Science Photo Library

p. 12
Science Photo Library

p. 14
All: History of Science Collections, University of Oklahoma Libraries;
copyright the Board of Regents of the University of Oklahoma

p. 15
All: History of Science Collections, University of Oklahoma Libraries;
copyright the Board of Regents of the University of Oklahoma

P. 16
All: History of Science Collections, University of Oklahoma Libraries;
copyright the Board of Regents of the University of Oklahoma

p. 17
Andrew Dunn (CC-AS 2.0)

p. 18-19
The Boerhaave Museum

p. 21
Sheila Terry/Science Photo Library

p. 22
History of Science Collections, University of Oklahoma Libraries;
copyright the Board of Regents of the University of Oklahoma

p. 23
Royal Astronomical Society/Science Photo Library

p. 24
(Top): Science Photo Library
(Bottom): Hubble Space Telescope/NASA, ESA (S. Beckwith (STScI), and
the Hubble Heritage Team STScI/AURA))

p. 25
(Left): Royal Astronomical Society/Science Photo Library. (Right): ESO

p. 26
Matthew Hunt; GNU Free Documentation License

p. 29
Alain Riazuelo

p. 30
NOAO/AURA/NSF

p. 31
Bill Miller

P. 32
Pedro Ré, George Willis Ritchey

p. 34
5-metre Hale Telescope's Wide-field Infra-Red Camera/California
Institute of Technology (David Thompson)

p. 35
Davide De Martin (ESA/Hubble), the ESA/ESO/NASA Photoshop FITS
Liberator & Digitized Sky Survey

p. 36
The Franklin Institute

p. 37
5-metre Hale Telescope with its Wide-field Infra-Red Camera/Califor-
nia Institute of Technology (David Thompson)

p. 38
Tom Jarrett

p. 40
Special Astrophysical Observatory (SAO)

p. 41
All: ESA/Hubble (M. Kornmesser)

p. 42
MPIA

p. 43
NOAO/AURA/NSF (Mark Hanna)

p. 44
ESO (C. Madsen)

p. 47
New Technology Telescope/ESO (Leonardo Testi and Leonardo Vanzi)

p. 48
(left) Blanco 4-meter Telescope/Cerro Tololo Inter-American Observatory (M. Robberto). (right) NOAO/AURA/NSF/Gemini Observatory

p. 49
SAGEM/ESO

p. 50
ESA/Hubble & ESO (Luis Calcada)

p. 53
ESO (H. Heyer)

p. 54
ESO (Yuri Beletsky)

p. 56
Gemini Observatory/AURA/NSF (Richard Wainscoat)

p. 57
Gemini Observatory (K. Pu'uohau-Pummill)

p. 58
NASA/JPL-Caltech

p. 59
NSO/AURA/NSF (Bill Livingston)

p. 60-61
Large Binocular Telescope Corporation (United States, Italy and
Germany, Marc-Andre Besel and Wiphu Rujopakarn)

p. 62
Hubble Space Telescope/NASA, ESA (M. Robberto (Space Telescope
Science Institute/ESA) and the Hubble Space Telescope Orion Treasury
Project Team)

p. 65
(Top): NASA & Flammarion, La Planète Mars
(Bottom): John Draper

p. 66
Pedro Ré, George Willis Ritchey

p. 67
Palomar Samuel Oschin Schmidt Telescope

p. 69
ESA/Hubble (M. Kornmesser)

p. 71
University of Alaska Anchorage (T.A. Rector), WIYN (H. Schweiker) and
NOAO/AURA/NSF

p. 72
Fermilab Visual Media Services

p. 74
LSST Corporation, Mason Productions (Todd Mason)

p. 76
NASA/JPL-Caltech/P.S. Teixeira (Center for Astrophysics)

p. 79
Adriaan Renting

p. 80
David Smyth

p. 81
NRAO/AUI (Kristal Armendari)

p. 82
Very Large Array Telescope (NRAO/AUI)

p. 83
Spitzer Space Telescope (NASA/JPL-Caltech)

p. 85
ESO (L. Calcada/H. Heyer)

p. 87
MAGIC Telescope Project (Robert Wagner, MPI)

p. 88
Observatorio Astronómico de La Plata (Guillermo E. Sierra)

p. 91
Coast Guard BM Photography (CC by-nc-nd)

p. 92
ESO

p. 93
Nik Szymanek

p. 94
NASA

p. 97
Hubble Space Telescope/NASA, ESA and A. Nota (ESA/STScI,
STScI/AURA)

p. 98
Hubble Space Telescope/NASA, ESA, and S. Beckwith (STScI) and
the HUDF Team

p. 99
Hubble Space Telescope/Robert Williams and the Hubble Deep
Field Team (STScI) and NASA/ESA

p. 101
Russ Underwood, Lockheed Martin Space Systems

p. 103
Chandra X-ray Observatory/NASA/UMass/D.Wang et al.

p. 105
NASA's Galaxy Evolution Explorer and the National Science Foundation's Very Large Array/NASA/JPL-Caltech/VLA/MPIA

p. 106
Infrared: Spitzer Space Telescope/NASA/JPL-Caltech/R. Kennicutt
(University of Arizona), and the SINGS Team; Visible: Hubble
Space Telescope/Hubble Heritage Team

p. 107
Hubble Space Telescope, Chandra X-Ray Telescope and Spitzer
Space Telescope, NASA, ESA, CXC, JPL-Caltech, J. Hester and A. Loll
(Arizona State Univ.), R. Gehrz (Univ. Minn.), and STScI

P. 109
WMAP/NASA (the WMAP Science Team)

p. 110
ESO (H. Zodet)

p. 112
Giant Magellan Telescope/Carnegie Observatories

p. 114-115
Thirty-Meter Telescope/Caltech, University of California (UC)
and the Association of Canadian Universities for Research in
Astronomy (ACURA)

p. 116
ESO (H. Zodet)

p. 117
Graeme L. White & Glen Cozens (James Cook University)

p. 119
SKA Project Office and XILOSTUDIOS

p. 121
JWST/NASA (Northup Grumman, ESA & CSA)

p. 122
ESA/Hubble (M. Kornmesser)

p. 123
Hubble Space Telescope/NASA, ESA (M.J. Jee and H. Ford (Johns
Hopkins University))

P. 125
ESA (Medialab)

p. 126-127
Hubble Space Telescope/NASA, ESA

p. 129
Babak Tafreshi/TWAN (twanight.org)

p. 130-131
(left) Hans Hordijk, (right) Bob Fosbury

p. 133
Hubble Space Telescope/ESA, NASA

Acknowledgements

We are grateful to Laura Simurda for painstaking picture research and editing, to Anne Rhodes for precise proof reading and to the dy-
namic design duo Martin Kornmesser and Nuno Marques for the nice layout and design of the book as well as for creating new illustrations
and doing image processing. We are also thankful to the many individuals who have provided images and advice along the way.